ENERGY EFFICIENCY
The Policy Agenda for the 1990s

The Policy Studies Institute (PSI) is Britain's leading independent research organisation undertaking studies of economic, industrial and social policy, and the workings of political institutions.

PSI is a registered charity, run on a non-profit basis, and is not associated with any political party, pressure group or commercial interest.

PSI attaches great importance to covering a wide range of subject areas with its multi-disciplinary approach. The institute's 30+ researchers are organised in terms which currently cover the following programmes:

Family Finances
Health Studies and Social Care
Innovation and New Technology
Quality of Life and the Environment
Social Justice and Social Order
Employment Studies
Arts and Cultural Studies
Information Policy
Education

This publication arises from the Quality of Life and the Environment programme and is one of over 30 publications made available by the Institute each year.

Information about the work of the PSI, and a catalogue of available books can be obtained from:

Marketing Department, PSI
100 Park Village East, London NW1 3SR

ENERGY EFFICIENCY
The Policy Agenda for the 1990s

Edited by
Ian Christie and Neil Ritchie

Policy Studies Institute
in association with
Neighbourhood Energy Action

The publishing imprint of the independent
POLICY STUDIES INSTITUTE
100 Park Village East, London NW1 3SR
Telephone: 071-387 2171; Fax: 071-388 0914

ISBN 0 85374 543 9

A CIP catalogue record of this book is available from the British Library.

PSI Report No. 736

How to obtain PSI publications
All book shop and individual orders should be sent to PSI's distributors:

BEBC Ltd
9 Albion Close, Parkstone, Poole, Dorset, BH12 2YL

Books will normally be despatched in 24 hours. Cheques should be made payable to BEBC Ltd

Credit card and telephone/fax orders may be placed on the following freephone numbers:

FREEPHONE: 0800 262260 FREEFAX: 0800 2622266

Laserset by Policy Studies Institute
Printed in Great Britain by Billing and Sons, Worcester

Contents

Foreword

The series of seminars on which this book is based were arranged to mark the 10th anniversary of Neighbourhood Energy Action. Whilst other events highlighted achievements and celebrated past successes, it is an indication of NEA's coming of age that it also wanted to look forward to the next decade and the search for a more effective solution to fuel poverty.

The origins of NEA lie in the traditional voluntary sector virtues of concern for disadvantaged people and an energetic and innovative response to a significant social problem. The commitment to provide a practical response to the problem of fuel poverty remains a distinguishing characteristic of NEA's work, and it has demonstrated both dedication and ingenuity in ensuring that people on low incomes benefit from energy efficiency improvements which can help them to keep warm and reduce fuel bills.

However, as with any organisation that grows and matures, it is inevitable that attention should also begin to focus on the wider policies that are needed to guarantee affordable warmth. It was also natural for NEA to invite the Policy Studies Institute to collaborate on its wider agenda. PSI is known for the strength of its research on social security and family finances, employment, energy and the environment and has an established Lunchtime Seminar programme that provides an invaluable independent forum for discussions of public policy. In commissioning background papers from acknowledged experts in the field and inviting distinguished speakers to address a selected audience of key decision makers, NEA and PSI have made an important contribution to the policy debate. The seminars were both timely and thought-provoking, since energy efficiency has dramatically re-emerged on to the political agenda in the wake of revelations about the environmental impact of energy consumption. Although this may be the catalyst for renewed interest, energy efficiency also has important social and economic benefits which should not be overlooked. I would like to thank all those who participated in the seminars, and who ensured that such a full analysis of all of these issues was provided, and to acknowledge in particular the important contributions made by the authors of the papers, the speakers, those who chaired the seminars, and the guests whose comments and questions ensured a lively and thought-provoking debate.

Whilst the individual seminar reports identify these key contributions, I would like to add a word of congratulation to the staff of NEA and PSI for their work in planning and organising the seminars. Finally, I would like to thank BP, whose sponsorship was crucial in enabling such an important debate about the future direction of energy policy to take place.

Lord Ezra
President, Neighbourhood Energy Action

British Petroleum
BP was delighted to support this series of seminars in order to assist the debate on this increasingly topical subject.

BP's commitment to energy conservation and effciency is explicitly stated in its Health, Safety and the Environment strategy. The three main reasons for our involvement are:

- BP is strongly pledged to protecting the environment.
- The attractions of making cost reductions through improved energy efficiency make obvious sense in the context of BP's own housekeeping.
- BP regards properly functioning markets as the best way to foster economic growth and therefore supports any efficiency measures which improve market mechanisms.

We wish this report every success in taking the issues raised to a wider audience.

Robert Horton
Chairman

Introduction

The revival of the debate on energy efficiency
This book is based on a seminar series held in the autumn of 1991 to mark the tenth anniversary of Neighbourhood Energy Action (NEA). In its first decade NEA has emphasised continually that the issues of energy efficency and fuel poverty should remain an important item on the agenda for politicians and the business community. However, the greater attention paid to energy efficiency and conservation in the 1970s was only a temporary consequence of the 'oil shocks' of 1973/74 and 1979, which concentrated Western minds on the security of oil supplies and the need to reduce dependence on imported fuel. This effect did not last long. The exploitation of North Sea oil and gas reserves and the relative fall in oil prices combined to make energy saving seem far from urgent to government, consumers and businesses in Britain during the 1980s. By 1991 the effects of the apathy of the eighties were apparent:

> The last few years have been marked by cuts in energy efficiency grants to industry and to householders, by increased VAT on energy efficient goods such as insulation and efficient lightbulbs, and by a fall in the real cost of fuel. Consequently, although last year industrial and commercial output fell, CO_2 emissions actually rose. In other words, the UK became even less energy efficient[1].

As NEA enters its second decade, however, the issues of energy efficiency and low income households' ability to afford light and warmth are steadily rising up the political agenda. It is clear that the 1990s will see an intensified debate on energy efficiency, but in a much-changed context from that of the 1970s. The renewed attention now being paid to energy efficiency and conservation is a consequence of the growing awareness of global environmental threats from the 'Greenhouse Effect' – the potentially disastrous warming of the Earth caused in large part by the emission of gases such as carbon dioxide as a by-product of our consumption of fossil fuels.

This book provides a guide to the new policy agenda for energy efficiency in the 1990s, and a stimulus to further debate on specific measures. It contains four papers commissioned as background material for the speakers and audiences at the seminars organised by NEA in conjunction with the Policy Studies Institute; and it summarises the contributions from the various distinguished speakers and guests at the seminars.

The first paper is on the environmental dimensions of energy efficiency, reflecting the key role ecological concerns now play in the development of energy policy. It is noteworthy that the Energy Efficiency Office was moved after the 1992 General Election to the Department of the Environment, while the rest of the wound-up Energy Department was absorbed into the Trade & Industry Department. The focus on global warming has, however, tended to obscure other important factors in the debate on energy efficiency, and these are explored in the other papers: the economic policy framework for energy supply within which a strategy for energy saving can be developed; the need to reconcile overall reductions in fossil fuel consumption with the fact that millions of households are unable to obtain sufficient heat, light and other energy services because of low incomes, poor insulation and inadequate heating systems; and the potential impact of major energy efficiency initiatives in providing new employment opportunities.

The advantages of energy efficiency investment

It was noted throughout the seminar series that energy efficiency seems to offer enormous benefits at minimal cost: as David Gee put it, it is more even than a 'free lunch' – it is 'a lunch you are paid to eat'. Given the economic savings and the environmental and social benefits in the form of reduced emissions, reduced fuel poverty and increased employment, many speakers expressed surprise and disappointment that governments, consumers and business were not rushing to take advantage. The potential rewards from investing on a large scale in energy efficiency are noted by all of the authors in this volume. The arguments in favour are compelling:

> The European Commission and the British Energy and Environment Ministers all agree that energy efficiency is the quickest and most cost-effective way to combat global warming...The potential for energy saving is enormous. The Government's own figures suggest cost savings of around £10 billion a year from the national fuel bill of £50 billion, with technically possible savings of up to £25 billion. The cost-effective savings alone would reduce the UK's current CO_2 emissions by

up to one-fifth, as well as reducing fuel bills, creating jobs and increasing the warmth of British homes(2).

As Tim Jackson observes in his paper on the environmental dimension of energy efficiency, energy savings provide not only a safeguard against the threats posed by rising energy consumption, but also offer an acceptable and readily available means of reducing dependence on fossil fuels. Renewable energy sources such as solar power and wind energy are not yet technically or economically competitive with established fuels: although this situation may change considerably by the turn of the century, the timetable of global warming demands preventive action as soon as possible. The development of nuclear power has been halted by deep-seated public fears over safety and waste disposal, and by the immense costs associated with it. The combination of unpopularity and expense makes nuclear power unattractive as an alternative to fossil fuels despite its lack of CO_2 generation. Against this background, investment in energy conservation and efficiency offers the quickest and least contentious route to reduced CO_2 emissions.

The environmental benefits from reduced pollution and above all from mitigation or avoidance of the threat of global warming are largely intangible: they relate to preventive measures over the long term. By contrast, the direct economic benefits from energy saving are more tangible, and the potential is remarkable. The figure of £10 billion in savings from adoption of best available techniques for energy efficiency and conservation is equivalent to the UK's annual income from North Sea oil(3). As Linda Taylor's review of the potential employment impacts of energy efficiency programmes indicates, there could also be great scope for job creation over a long timescale.

Finally, there are advantages for social policy from increased energy saving. NEA's goal is to combat fuel poverty in the UK by improving energy efficiency in low-income households, and the debate on energy efficency cannot ignore the social dimensions of energy policy. As Brenda Boardman notes in her paper, nearly one-third of British households cannot afford adequate warmth on their present income: over 7 million households are in 'fuel poverty', and almost all of them live in poorly-insulated homes. Given the environmental imperative to reduce overall fuel consumption, how can those in fuel poverty be provided with adequate warmth and lighting? As Brenda Boardman and other participants in the seminar series observed, energy pricing measures to combat global warming – such as a carbon tax – could make matters worse for low-income households. The twin policy objectives of cutting CO_2 emissions and providing affordable warmth to all can only be reconciled by investing in energy efficiency, starting with the worst-insulated low-income households.

Constraints on investment in energy efficiency

Much of the debate in the seminars, as well as much of the analysis in the papers collected in this volume, centred on the existing constraints on energy efficiency investment that prevent the benefits from being realised.

A key issue is the price of energy. With a decline in the real cost of fuel over the past decade, there is little incentive to domestic or business consumers to make investments in energy saving. However, although higher energy prices seem inescapable if progress is to be made on reducing fossil fuel emissions, there is great controversy about how to implement such price rises. As several contributors to the seminars noted, a carbon tax or other energy levy would need to be set at a very high level to make any impact on demand; and across-the-board energy taxes would have regressive effects on low-income households. This alone indicates that the design and implementation of a carbon tax within the European Community in general and the UK in particular will be the subject of a contentious and prolonged debate. The fears of governments and industry over a possible loss in international competitiveness because of the extra costs imposed by a carbon tax are also certain to constrain the development of policy on energy pricing for environmental protection.

Not only is there little consensus about the design of the stick that higher energy prices would represent, there is great disagreement about the carrots that could be offered to consumers to provide an incentive for investment in energy efficiency. This debate has to be set against the overall policy framework of the past decade. The UK Government's approach has been resolutely market-oriented and opposed to the development of a strategic energy policy (5). The Government has privatised the gas and non-nuclear electricity industries, reduced energy efficiency grants to business and households, and confined its role largely to regulating the energy market. In the energy efficiency sphere, the main emphasis has been on limited initiatives to improve insulation in low-income households and on publicity campaigns to encourage domestic and business consumers to save money by saving energy.

A constant theme of the discussion in the seminar series was the limitations of the market-oriented approach in promoting energy efficiency. There was considerable concern about the lack of a comprehensive, integrated strategy from government for securing energy efficiency improvements and correcting market failures in this regard. Critics argued that for many years UK energy policy had been focussed almost exclusively on the supply side, with inadequate attention paid to demand side issues. Numerous criticisms were made of what was seen as the weak position of the Energy Efficiency Office within Whitehall, lack of attention to least-cost planning on the supply side, insufficient energy efficiency standards for buildings and appliances, and the failure of public agencies to set a good example in making their own buildings energy efficient.

The natural emphasis in the privatised energy supply companies is on increasing sales rather than on promoting measures that would reduce demand. It was frequently argued that the utilities should be given more incentives to sell 'energy services' rather than simply units of power, and that tariff structures should be revised to make it attractive to invest in energy efficiency improvements. The gas industry, through the new 'E-factor' formula, is to have such an incentive built into its tariff structure from 1992. Such a measure has not yet been introduced for the electricity industry.

In this context many speakers raised the issue of 'Least Cost Planning' (LCP) techniques, which have developed significantly in the United States, in many cases as a result of legislative pressure on the energy utilities. LCP covers a variety of strategies to minimise the need for further investment in generating capacity and to maximise the scope for cost savings on the part of energy suppliers and consumers. These include developing markets in saved electricity ('negawatts') and making electricity savings tradeable between generating companies. Several speakers called for more attention to LCP techniques in the UK to increase energy efficiency and stimulate suppliers to become providers of energy services. As John Chesshire notes in his paper, this debate has been affected by privatisation of the gas and electricity utilities: government tends to regard discussion of LCP and other aspects of the balance between supply-side and demand-side management in energy as matters for the utilities in consultation with the regulators Ofgas and Offer. Yet many contributors to our seminar discussions saw no alternative to a more strategic role from government in stimulating a market for energy services in which LCP techniques would play a much greater role than at present.

Many contributors commented on the prevalence of shortcomings in information provision about energy saving, which fundamentally affected the development of the market for energy efficiency products and services. The lack of consumer awareness and of useful advice to households and businesses on energy saving was frequently noted. Research for the Department of the Environment indicates that many consumers are ill-informed about the costs and benefits of energy efficiency, do not see it as a priority or a financially worthwhile investment, and fail to make the connection between energy consumption and environmental problems(4). It was argued in the seminars that the energy suppliers were poor at providing information and advice on energy costs and efficiency to consumers. There were numerous calls for better information to consumers in the form of energy labelling of products and houses, and for cuts in VAT on energy-efficient products and energy-saving materials to stimulate demand. In the absence of any intervention in the market, the rate of insulation-fitting in the housing stock, and of take-up of low energy appliances, is likely to be disastrously slow:

> On present demand and without any stimulus, cavity wall insulation is expected to take another 50 years to grow from the present 20 per cent penetration of the market to 70 per cent. Condensing boilers are expected to take 150 years and low-energy lights 220 years to reach a 20 per cent take-up(6).

An agenda for the 1990s

On the basis of the papers and seminar discussion, and in relation to recent policy developments in the UK and European Community, it seems clear that a new agenda for energy efficiency is emerging. Regardless of the political complexion of governments through the 1990s, it is evident that environmental pressures will cause more attention to be paid to long range strategic issues in energy policy. Given the economic and technical constraints on alternatives to fossil fuels, energy efficiency and conservation are now recognised across the political spectrum as the best routes to reduced CO_2 emissions. Whilst there is no consensus yet on the most effective means of ensuring greater investment in energy efficiency, it seems likely that the following will become key elements of energy policy in the coming years:

– Improved information for domestic and business consumers through energy labelling and audits, and incentives for insulation measures by householders, to stimulate the market for energy-efficient products;

– Incentives for the energy suppliers to offer services in energy management and energy advice, and to invest in energy efficiency measures, including Least Cost Planning techniques;
– A higher profile for energy efficiency policy in government: the Advisory Committee for Business and the Environment set up in May 1991 by the Secretaries of State for Environment and Trade and Industry has recommended a wide range of government measures to promote energy efficiency and raise standards, including a stronger role for the Energy Efficiency Office(7). In addition, policy making in energy at the European Community level is certain to become ever more significant, and Brussels is likely to be the source of most new initiatives on reducing CO_2 emissions and tightening energy efficiency standards;
– Increased targeting of insulation measures on low-income households to mitigate fuel poverty and cut energy loss from the most poorly insulated homes. Already the Government has announced a rise of some 50 per cent in the budget of the Home Energy Efficiency Scheme (HEES), allowing 250,000 homes to be given insulation treatment in 1992/93.

There is, then, an emerging agenda for energy efficiency policy that stresses a more strategic approach than has been fashionable for the past decade. This new policy agenda is being shaped principally by environmental pressures, and is likely to be as much influenced by the European Commission as by Whitehall, if not more so. The potential benefits of an integrated approach to energy efficiency are substantial, as the papers collected here make clear: large financial savings; large gains in employment; a reduction in fuel poverty; and a major contribution to diminishing the threat of global warming.

Format of the book
The first paper, by Tim Jackson, provides an overview of the environmental dimensions of energy policy and of the environmental benefits that can flow from a comprehensive strategy for promoting energy efficiency and conservation. Energy efficiency measures offer the most readily available and reliable means of reducing CO_2 emissions and mitigating the Greenhouse Effect.

John Chesshire's paper examines the economic aspects of energy efficiency policy, concentrating on the neglect of demand-side issues in UK energy policy and the limitations of a market-oriented approach to energy in promoting greater efficiency and conservation. The paper includes a list of key issues for the 1990s, including more attention to least-cost planning, reliable information on energy use, a stronger role for the EEO, and a careful balance between the use of taxes and subsidies to encourage energy saving.

Brenda Boardman considers the social aspects of energy efficiency policy, focussing on the extent of fuel poverty in the UK and the policy constraints that affect its alleviation. She sets out a long term programme to bring 'affordable warmth' to low income households at the same time as promoting energy savings, through a major government-led capital investment initiative.

Finally, Linda Taylor reviews recent studies of the employment impacts of large-scale initiatives for investment in energy saving. Whilst there is no consensus on the costs of such programmes, the potential for considerable job creation alongside savings in energy and reduced pollution is clear.

Each paper is followed by a summary of the proceedings of the relevant seminar. The summaries include a synopsis of the presentation by the speaker and a brief account of the ensuing discussion.

Ian Christie
Policy Studies Institute

Neil Ritchie
Neighbourhood Energy Action

References

1. Association for the Conservation of Energy, *First Steps: towards a national energy efficiency programme*, ACE, London, November 1991.
2. ibid. For detailed analysis of technologies for efficient energy use in relation to the issue of global warming, see Michael Grubb et al, *Energy Policies and the Greenhouse Effect. Volume 2: Country Studies and Technical Options*, RIIA/Dartmouth, Aldershot, 1991, chapter 2.
3. David Young, 'Energy campaign reaches boiling point', in *The Times*, 20 January 1992.
4. A. Hedges, *Attitudes to Energy Conservation in the Home*, HMSO, London, 1991.
5. See Frances McGowan, 'UK Energy Policy', in *ENER Bulletin*, European Network for Energy Economics Research, July 1991.
6. Pearce Wright, 'Economy begins at home', in *The Times*, 20 January 1992.
7. Advisory Committee on Business and the Environment, *First Progress Report to and Response from the Secretaries of State for the Environment and for Trade and Industry*, DTI/DoE, October 1991.

The environmental benefits of energy efficiency

Dr Tim Jackson
Centre for Environmental Change
Lancaster University

Introduction

The physicist Boltzmann once described the struggle for existence as the struggle for free energy. Twentieth century human society has - to a large extent - been ignorant of that struggle because of the availability of energy carriers such as fossil fuels (coal, oil, gas), and nuclear power. Fossil fuels, in particular, have facilitated a dramatic increase in the energy intensity of civilisations since the industrial revolution, and unshackled mankind from energy limitations. The consequences of that freedom are evident in our increased mobility, the reduced burden of manual work, generally increased longevity, and the increased complexity of our industrial and social infrastructure.

Nevertheless, this freedom has not been without a price: the shadow of environmental degradation. Increased concern for the protection of the natural environment is now a fact of modern political life. This growing awareness has particular implications for energy policy. Many of the environmental problems facing the world today are related, directly or indirectly to energy consumption.

Acid rain and global warming are well-known examples. There are a host of other problems however. These include the emission of particulate matter (smoke), unburnt fuel (hydrocarbons), toxic metals present as trace contaminants in fossil fuels, methane gas, chlorine, and ozone; as well as the deterioration of the natural landscape. The environmental problems associated with energy production range from visual intrusion to potentially catastrophic global environmental change. Power plant siting, for example, leads to visual impacts.

Mining and drilling activities produce disruptive effects on local ecosystems. Toxic health effects on humans arise from heavy metals and unburnt hydrocarbons. Global environmental effects include forest dieback, acidification of lakes and soils, sea-level rise, increased desertification and climatic instability.

In the light of these emerging difficulties, increased attention is being given to the virtues of energy conservation, that is, reductions in the use of energy resources. Particular attention is being given to energy efficiency, which can lead to reductions in use without loss of service. The efficient use of resources has several clear advantages. It will slow down the decline (in quantity and in quality) of such resources, and thereby improve their availability to future generations; one of the earliest reasons for interest in energy efficiency was concern about resource scarcity. Efficiency of resource use also has economic implications; the oil-price shocks of 1973/4 and 1979, provided sharp incentives for an increased concern for energy efficiency. Apart from implications of economics and scarcity, the various stages of resource use - extraction, refinement, conversion, consumption, and dissipation - all have significant environmental impacts. Efficiency of utilisation of resources will reduce those impacts. This is as true of energy as it is of any other resources. The importance of energy efficiency is no better illustrated than by the increasing awareness and concern over the man-made greenhouse effect. Potentially catastrophic global changes may be in prospect, largely as a result of the energy consumption patterns of industrial economies.

The environmental benefits of the efficient use of resources are clearly determined by the environmental impacts of the use of those resources. This paper is therefore concerned with describing what those impacts are in the case of energy resources. It concentrates on energy consumption in the UK domestic sector. The next section gives a broad outline of energy consumption in the UK, particularly in the domestic sector, determining the demands for primary fuel, and the end-uses of consumption. We then examine the environmental impacts associated with this pattern of energy usage. A brief description follows of the various possibilities for energy efficiency, particularly in the UK domestic sector. Finally, the policy aspects involved in the implementation of energy efficiency are considered.

Patterns of energy consumption in the UK (1)

Energy demand in the UK in 1989 was 6,230 PJ (petajoules)[*]. Twenty seven per cent of this was the delivered energy demand of the domestic sector (Figure 1). This sum represents energy delivered to end-users (homes, business, industry, agriculture etc). Because of losses in conversion, transmission and distribution the total primary fuel consumption is always substantially bigger than the delivered energy. In 1989, primary fuel consumption was 8,930 PJ. By far the greatest part of this difference between energy demand and primary fuel consumption results from losses incurred in electricity generation. Conventional thermal electricity generation has an efficiency of around 35 per cent. Almost two thirds of the energy consumed in thermal power stations is therefore lost.

Assuming that conversion and distribution losses for each delivered fuel type are equally apportioned between the different sectors, the proportion of primary fuel consumed by the domestic sector was 29 per cent (Figure 2). Primary fuel consumption by the domestic sector is a slightly higher proportion of total fuel consumption than of energy demand, because of the high proportion (relative to other sectors) of electricity used in the domestic sector.

The conceptual and quantitative difference between primary fuel consumption and delivered energy is of considerable interest in understanding the potential for improved energy efficiency. As already remarked, the difference between the two figures arises from the losses which are incurred in conversion, transmission and distribution. Associated with these losses, of course, are considerable environmental impacts. It is certainly pertinent to ask, therefore, whether it is possible to reduce those losses that is, whether it is possible to improve the efficiency of supply of energy. We can refer to this as *supply-side energy efficiency.*

Such improvements would include, the minimisation of leaks, and accidental losses, the development of better transmission and distribution systems and the use of improved-efficiency conversion systems. We discuss the technological possibilities for such improvements later.

---

[*]One of the difficulties of making sense of energy demand and supply is that a number of different units are employed. Different units are commonly used for different fuel types (for instance, units for gas consumption are commonly measured in therms whilst, units of electricity consumption are commonly measured in kilowatt hours), and sometimes according to whether end use or primary fuel is being discussed. In this paper I shall use petajoules (PJ) throughout: 1 petajoule = 10^{15} joules.

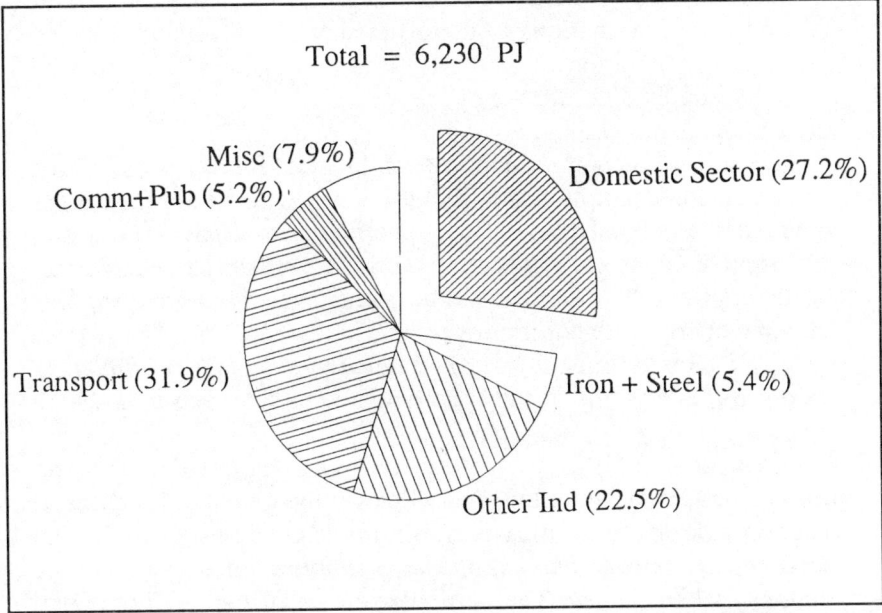

Figure 1: Demand for Delivered Energy by Sector (UK 1989)

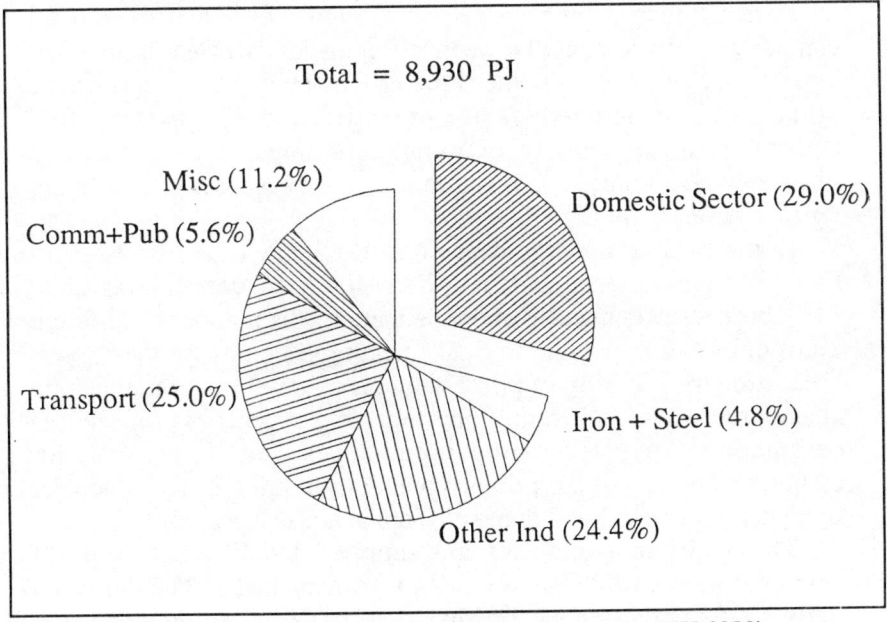

Figure 2: Primary Fuel Consumption by Sector (UK 1989)

Here we remark only that - despite the existence of certain thermodynamic limitations to the efficiency of conversion systems (particularly generation of electricity) - there is in general considerable scope for supply-side efficiency measures to reduce the difference between the demand for energy and the primary fuel consumption.

The second area in which we might look for ways to reduce primary fuel consumption is in the efficiency with which final users convert delivered fuel (coal, oil, gas, electricity) into *services* (warmth, light, motive power etc). It is energy services rather than energy that consumers want. The provision of the same level of service using less delivered energy calls for improvements in the efficiency of final use of energy. This is sometimes called *demand-side energy efficiency*. We discuss later some of the specific demand side measures which can be adopted.

Irrespective of the differences between energy demand and primary energy consumption, and whatever the potential for demand or supply side efficiency measures, we should not lose sight of the fact that the *environmental impacts of energy supply are determined by the primary fuel consumed*. The particular nature of the environmental impacts will depend on the type of primary fuel consumed. Impacts associated with coal burning, for example, are of a completely different nature to the impacts associated with nuclear power. Making comparative distinctions between different kinds of environmental impact is notoriously difficult. Pure efficiency improvements on the other hand, which reduce the overall need for primary fuel consumption, are clearly of paramount importance. Figure 3 illustrates the primary fuel consumption in 1989 according to different types of primary fuels.

Domestic sector energy demand in 1989 was 1,700 PJ. Primary fuel consumption accounted for by domestic sector demand was 2,590 PJ. This energy demand was divided amongst a number of different kinds of *end-uses* (Figure 4). Space heating accounts for the biggest proportion (54 per cent) of primary fuel consumption. Water heating also represents a significant proportion (18 per cent) of primary fuel consumption. A smaller proportion (8 per cent) is required for cooking. The rest of the primary fuel consumption is accounted for by lights (3 per cent) and domestic appliances (17 per cent).

These different end-uses are supplied by different kinds of delivered fuels and different kinds of primary fuels. The delivered fuels are essentially coal, oil, gas or electricity. Space and water heating is now largely supplied by natural gas, although there is also some coal-fired, oil-fired and electric heating. Cooking is either gas

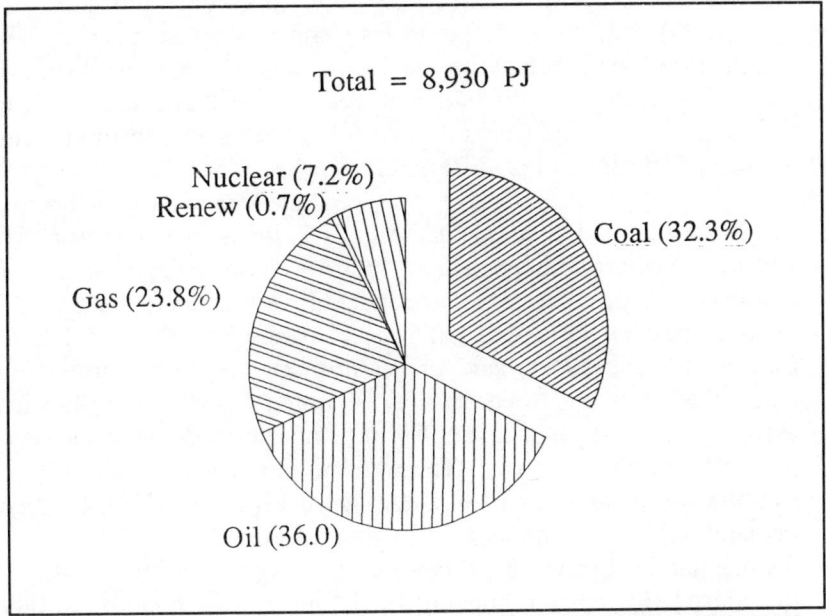

Figure 3: Primary Fuel Consumption by Primary Fuel Type (UK 1989)

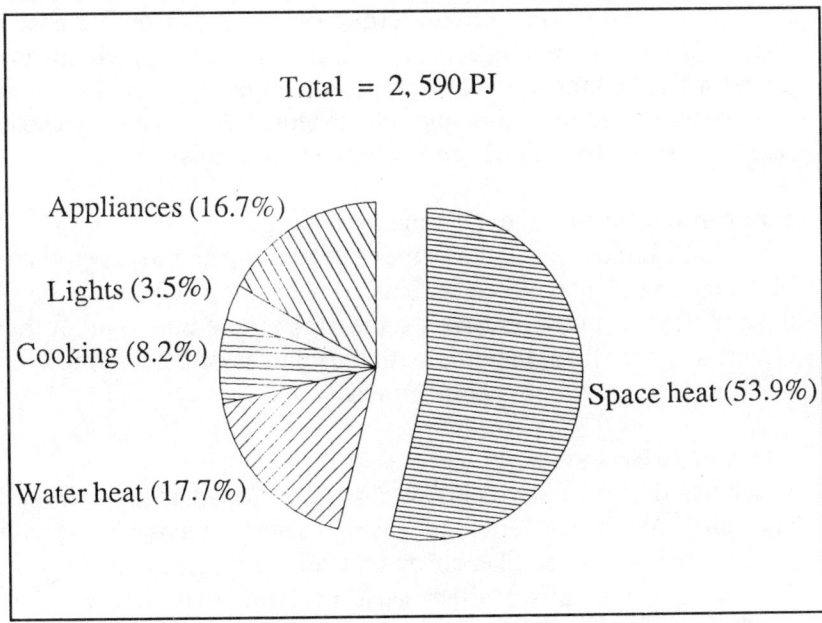

Figure 4: Domestic Primary Fuel Comsumption by End-Use (1989)

or electric. The primary fuels are essentially coal (and solid fuel derivatives), oil, gas, nuclear energy and renewable energy. For instance, primary fuels for electricity supply include coal, oil, renewable energy (mostly hydro-power) and nuclear energy. Figure 5 illustrates the relative proportions of domestic primary fuel consumption supplied by different fuel types.

This brief outline suffices to provide some idea of the way in which fuel is used in the UK today, particularly in the domestic sector. The question of what will happen in the future is amongst the most difficult questions it is possible to ask about energy demand and supply. It is certainly beyond the scope of this paper to present any kind of detailed forecast for energy demand or for primary fuel consumption. A number of different forecasts may be found elsewhere (2) but it is certainly true to say that the trend is of rising energy demand and rising primary fuel consumption. The trend in primary fuel consumption over the years 1985 to 1989 is shown in Figure 6. Future energy demand will depend on essentially two different kinds of factors: on the one hand activity-related factors (such as population, growth rate of industry, increase in housing stock, and so on), and on the other hand efficiency factors, including both supply side and demand side efficiency. Assuming efficiency remains more or less at 1989 levels, models of energy demand for 2005 tend to predict an increase of around 25 per cent in total energy consumption. (3) This will mean of course a similar increase in environmental impacts, except in so far as it is possible to apply end-of-pipe abatement technologies (which are relevant mainly for sulphur and nitrogen oxide emissions).

Environmental Impacts Of Energy Use

It would be impossible in the scope of this brief paper to categorise all of the environmental effects associated with energy consumption. In the next few sections, however, some of the more important of these effects are described, together with the sources of those effects, and the options available for abating them.

Particulate Emissions

When wood, coal or petroleum products are burned, most of the solid or liquid fuel is converted to gases. There remains some solid particulate matter or smoke, however, which is emitted with the gases. This solid matter arises either as a result of inefficiencies in the combustion process, or else because of impurities in the fuel. Coal, for example, contains approximately 17 per cent by weight of ash.

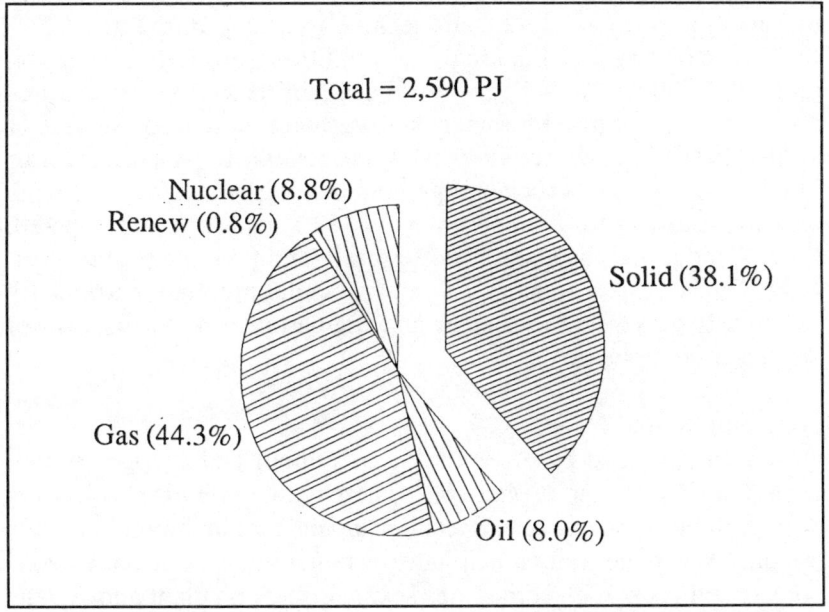

Figure 5: Domestic Primary Fuel Consumption by Fuel Type (1989)

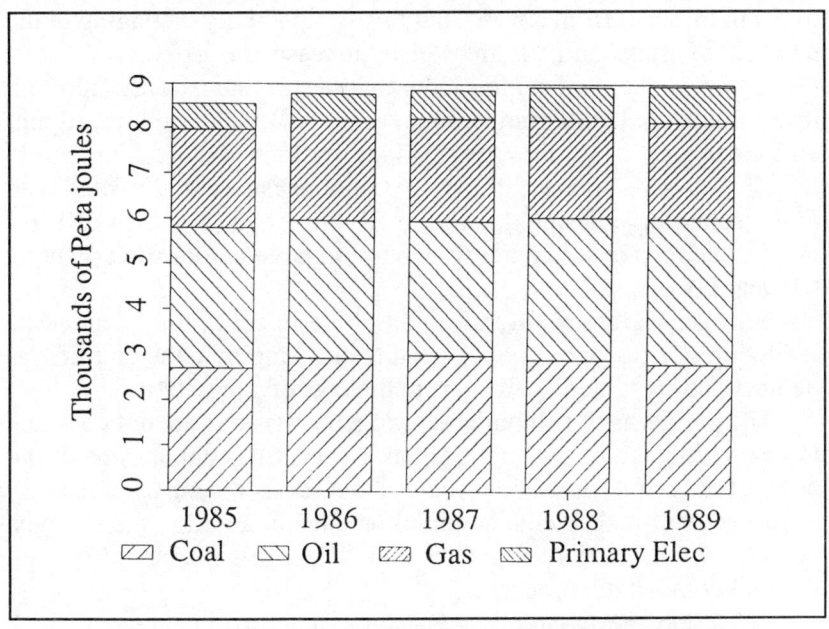

Figure 6: Recent Trends in Primary Fuel Consumption

Around 40 million tonnes of particulate matter is generated annually through the combustion of fossil fuels. About 34 per cent of this arises as a result of domestic sector energy consumption.

The severity of the local air pollution problem caused by particulate matter was largely mitigated by a series of measures introduced to control emissions, starting with the Clean Air Act of 1956. These measures certainly had some positive impacts on local air quality through the sixties and prevented a recurrence of the events such as the disastrous London smog of 1952. It should be remembered, however, that the disposal of bottom ash, and precipitated 'fly-ash' from the flue gases, remains an environmental problem, particularly as landfill sites become scarcer and regulations on the disposal of wastes at sea tighten.

Acid Emissions

Acid rain is caused by the emission of sulphur and nitrogen oxides emissions into the atmosphere (4). These oxides combine with water vapour in the atmosphere and result in an increase in the acidity of the vapour. When the water precipitates as rain, its acidity damages soils, forests and lakes with sometimes severe effects on plant and aquatic life.

One option for reducing sulphur emissions is to import low-sulphur coal from abroad, but this is potentially damaging to the balance of trade and it may also increase the effects of social disruption in the coal-mining communities. Aside from importing low-sulphur coal (or burning less coal), the only way to reduce sulphur emissions is to install flue gas desulphurisation (FGD) equipment. In the UK, however, none of the power stations has yet been fitted with FGD technology, despite the recent EC Directive (5) which calls for 60% reductions in sulphur emissions from large combustion plant by the year 2003.

Nitrogen oxides (NO_x) also contribute to acid rain. Large-scale reductions of nitrogen oxides require the employment of flue gas denitrification, which again adds to the cost of generation.

The problems of sulphur and nitrogen emissions are not confined to coal combustion. Natural gas has a very low sulphur content but heavy fuel oils (which contribute about 11 per cent to electricity supply) have a rather high sulphur content. In addition the nitrogen content of combustion air means that NO_x emissions arise from all thermal combustion processes.

In Europe, the so-called '30%-Club' - a group of nations of which Britain is not yet a member - has pledged to reduce sulphur emissions by 30 per cent over 1980 levels by the year 1993. In addition the signing of the 1988 EC Directive on acid emissions from large combustion plants and a new draft directive on vehicle emission levels have together done much to reduce public and political pressure about acid rain. This does not mean to say that the environmental problems themselves have been solved. Rather, concerns have been displaced by more recent pressures about the so-called 'Greenhouse Effect'.

Greenhouse Gas Emissions

The Greenhouse Effect is the warming of the global atmosphere as a result of increased concentrations of certain gases in the atmosphere. Although scientific consensus on the timing and regional effects of a potential global warming has yet to be achieved, little doubt now remains that such an effect will arise. Some global warming has already been observed, and increased concentrations of the 'greenhouse gases' have been measured in the atmosphere. The consequences of the global warming effect are also to some extent a matter for conjecture. Predicted consequences include sea-level rise, climatic instability, water shortages, increased desertification, and disruption of agricultural patterns.

The gases which contribute to the greenhouse effect include carbon dioxide, methane, nitrous oxide, ozone, and CFCs (chlorofluorocarbons). Carbon dioxide arises mostly from the emission of fossil fuels. Methane is the main constituent of natural gas, which is also an important energy carrier. Nitrous oxide from car exhausts is a significant source of potential warming. These facts underline the relevance of the greenhouse effect for energy policy. The main greenhouse gas is carbon dioxide (CO_2), which contributes about 50 per cent of the greenhouse effect (6). Particularly in the light of its inevitable link with fossil fuels, it is instructive to consider carbon dioxide emissions energy consumption in some detail.

Carbon Dioxide

Carbon dioxide is an inevitable product of burning fossil fuels. There is no practicable technology for removing carbon dioxide from flue gases. There is a limited potential for switching from one kind of fuel to another. The carbon content of different fossil fuels differs considerably. For example, natural gas releases less carbon for each unit of energy than oil, which releases less carbon for each unit of energy than coal (Table 1). This means that replacing coal-fired combustion by gas-fired combustion can reduce carbon dioxide emissions into the atmosphere.

Fuel	MtCO$_2$/TWh
Coal	0.31
Oil	0.26
Natural Gas	0.18

**Table 1: Carbon Coefficients for Different Fuels
(Million Tonnes of Carbon Dioxide per TeraWatt Hour)**

The only other options for reducing carbon dioxide emissions are either to reduce the energy generated from fossil fuel sources, by changing to renewable or to nuclear energy technologies, or to reduce the total consumption of energy. Reducing the amount of energy consumed can be done, either by reducing the level of services provided by energy, or else by improving the efficiency with which energy is used to provide those services. The threat of global warming, more perhaps than almost any other environmental impact of energy consumption, highlights the need for the efficient use of energy.

In response to the threat of global warming, there have been various calls for reductions in carbon dioxide (CO2) emissions. But only a limited number of countries have pledged themselves to actual reductions in CO2 emissions (see Table 2). The UK has announced its intention to 'stabilise' CO2 emissions at 1990 levels by the year 2005.

Total carbon dioxide emissions in the UK during 1989 were around 590 million tonnes. The domestic sector and the transport sector contributed around 27 per cent each to these emissions (Figure 7), underlining again the responsibility of the individual consumer in determining the environmental impacts of energy consumption. A little over half of the total domestic sector emissions of 158 million tonnes arose from space heating requirements (Figure 8). Electric lights and appliances also contributed a significant proportion.

Other Greenhouse Gases
The other greenhouse gases include methane, nitrous oxide, and ozone. Reduction in emissions of these gases also has implications for energy policy.

Man-made methane emissions in the energy sector arise from coal-mining and from the oil industry (7), and from leakage in the

Country	Goal
Australia	Reduce GHGs by 20% by 2005
Austria	Reduce CO_2 by 20% by 2005
Canada	Freeze CO_2 at 1990 level by 2000
Denmark	Reduce CO_2 by 20% by 2005; 50% by 2020-2040.
France	Freeze CO_2 near 1990 per capita level
Germany	Reduce CO_2 by 25% by 2005
Japan	Freeze CO_2 at 1990 level by 2000
Netherlands	Freeze CO_2 at 1990 level by 1995; then reduce.
New Zealand	Reduce CO_2 by 20% by 2005
Norway	Freeze CO_2 at 1989 level by 2000
Sweden	Freeze CO_2 at 1988 level by 2000
Switzerland	Reduce CO_2 10% by 2000
UK	Freeze CO_2 at 1990 level by 2005

Table 2: Announced Targets for Reductions in Greenhouse Gas Emissions (Source: Flavin, C. and Lenssen, N. *'Policies for a Solar Economy'* Energy Policy 1991, March 1992, Vol.20, no. 3.)

natural gas distribution and supply network (8). Estimates of gas leakage (as high as 10 per cent of gas supplied) have significant implications for global warming, since methane is considerably more powerful as a greenhouse gas than carbon dioxide. Very high methane leakage rates of gas could offset the advantage which natural gas has over coal in terms of carbon dioxide emissions. When methane emissions from coal mining are also taken into account, however, a clear advantage still remains with natural gas. Nevertheless, the

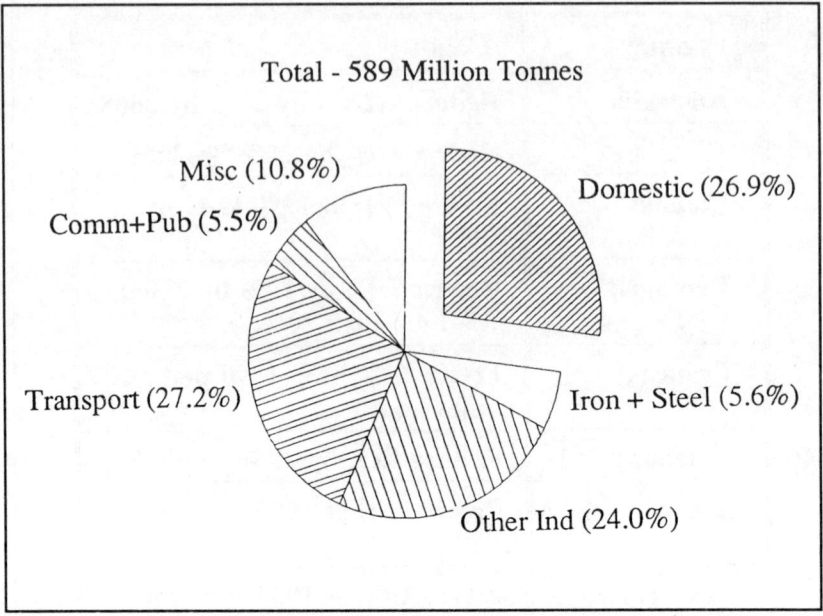

Figure 7: Carbon Dioxide Emissions by Sector (1989)

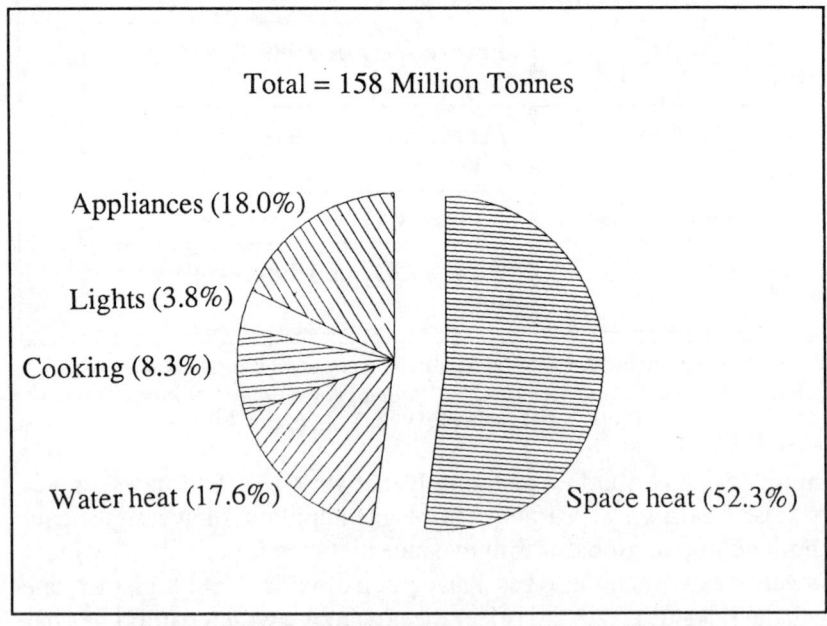

Figure 8: Domestic Sector CO$_2$ Emissions by Enduse

greatest benefits are obtainable by reducing fossil fuel consumption overall.

Nitrous oxides are formed during high-temperature combustion. Thermal energy conversion processes are therefore major culprits in the emission of nitrous oxide. Ozone in the lower atmosphere is formed through photochemical reactions between combustion exhaust gases, particularly from vehicle emissions. It is also formed electrostatically in the vicinity of high tension power lines.

In summary, the energy supply industries are heavily implicated in the greenhouse effect. Indeed 60 per cent of the greenhouse effect is believed to be related to man-made energy conversion and usage. Most of the contributions to the global warming arise from the combustion of fossil fuels, so that a reduction in fossil fuel usage is clearly crucial to future attempts to reduce the threat of global warming. Principally, these reductions must come from improvements in the efficiency of supply of energy and in the efficiency of use of energy.

There is some potential for switching from fossil to non-fossil energy supply options. Non-fossil energy supply options include nuclear power and renewable energies. The environmental effects of most renewable energies are comparatively few, and generally localised: for example, visual intrusion on the landscape from wind turbines. On the other hand, the potential for renewable energy technologies is limited by available technology, by land-use considerations, by local geographic and climatic conditions, and by economic factors (9). Nuclear technology is relatively well-developed, although increasingly expensive, and increasingly subject to public opposition, largely on account of its environmental impact.

Environmental Impacts of Nuclear Power

Environmental objections to nuclear power are well-known. They centre around the safety of nuclear plant, the routine emission of radioactivity into the working environment and into the immediate vicinity of the plant, the disposal of radioactive wastes (radwastes) and the decommissioning of nuclear power stations.

Accidents at Windscale in the UK, Three Mile Island in the US, and Chernobyl in the USSR have brought wide-spread attention to the potentially catastrophic impacts of nuclear power generation. There is a continuing debate about the health effects of the Chernobyl accident, with the nuclear industry arguing (10) that the accident had produced 'no health disorders that could be directly attributed to radiation 'exposures' and environmental lobbies claiming (11) that the science has been 'perverted by vested interests and PR agencies'.

In the United States, public concern over the lack of inherent safety, and private sector concern over rising costs in all elements of the nuclear cycle have virtually halted the construction of new nuclear plant. In the UK, a moratorium has been imposed by the government on the building of any new nuclear plants until 1994, when the programme will come under review.

Emissions of radioactivity from leaks and routine releases also amount to a continuing problem for regulators of the nuclear industry. One of the problems is the difficulty of determining the health effects of these releases. Although controlled radiation doses are used in medical treatment of cancers and tumours, uncontrolled radiation is generally hazardous both to human health and to the environment.

The problem of finding suitable repositories for all kinds of radwaste is one of the major uncertainties still hindering accurate assessment of the full costs of nuclear power. The two disposal routes for low-level waste favoured by the Central Electricity Generating Board in the UK (namely, sea-dumping and shallow burial) have been discarded on environmental grounds. This change in policy will mean a significant increase in radwaste disposal costs. The problem of locating an economically, socially and environmentally acceptable deep repository for low (LLW) and intermediate (ILW) wastes has still not been resolved. Deep-burial will be costly, and difficult to monitor. High level wastes (HLW) arise primarily either directly from spent fuel, or indirectly from reprocessing of spent fuel. There is as yet no high level waste disposal strategy in the United Kingdom. Since 1981 government policy has been that these wastes should be stored for up to 50 years.

Decommissioning is a further difficulty. As the first generation of nuclear power stations (the Magnox stations) approach the end of their useful life, the full costs, and safety implications of nuclear power are beginning to emerge. The problems of decommissioning are in part those of radioactive waste disposal. Quite apart from these problems however, there is the sheer physical difficulty of dismantling a highly radioactive structure without compromising workers' safety. Current plans include the contingency that the central core of the reactor should be buried in concrete for about a century in order to allow the radioactivity to cool down. The truth is however, that until a major nuclear facility has been successfully decommissioned, no-one will really know the full extent of the problems which will have to be solved.

The total annual waste arising from nuclear power generation in the UK is 31,000 cubic metres, of which 11,000 cubic metres is generated by domestic sector requirements (Figure 9).

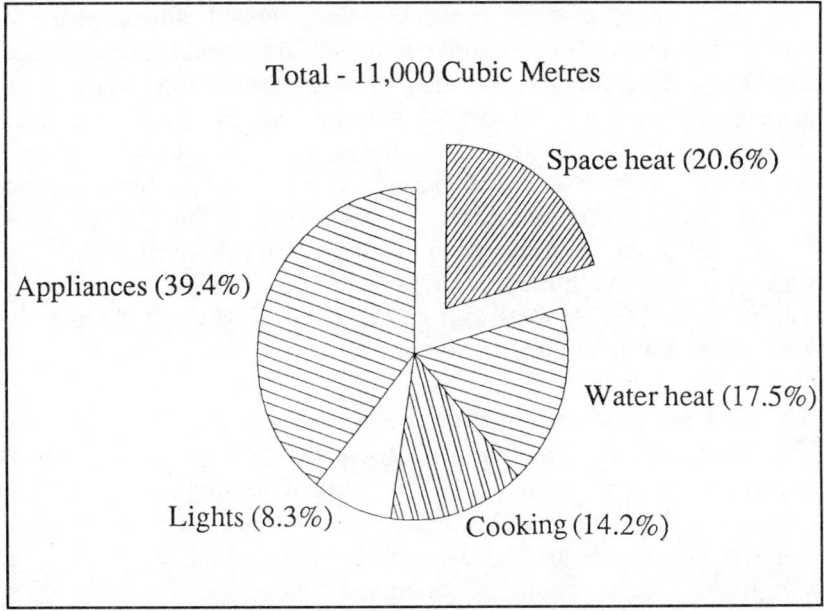

Total - 11,000 Cubic Metres

Space heat (20.6%)

Appliances (39.4%)

Water heat (17.5%)

Lights (8.3%)

Cooking (14.2%)

Figure 9: Annual Domestic Sector Radioactive Waste Arisings (1989)

Further environmental impacts

Acid emissions, greenhouse gas emissions and nuclear waste, are certainly the three best-publicised impacts of the energy supply industries. Each of these issues has reached the top of both national and international agendas, and each entails environmental effects which could have significant long-term and global impacts. There are however, a number of other lesser environmental concerns arising from the energy supply industries.

Land-use questions become particularly important as development increases. Apart from the visual intrusion on the land-scape, land used for energy supply industries cannot be used for agriculture, amenity, or for sustaining the health of ecological systems. Particularly important are the land-use requirements for mining. In some areas of Europe (including parts of the UK) the effects of open-cast mining are amongst the most striking impacts of energy production on the environment.

Hydro-Carbons. A further environmental problem is the release of hydro-carbons from petroleum based fuels and natural gas, either through incomplete combustion, or through evaporation, leakages and spills during distribution or supply. Once released into the environment these compounds contribute to the general toxic burden imposed by hydrocarbons on living beings. In humans, inhalation of hydrocarbons can promote respiratory diseases and cause cancer.

Trace elements are released when fuels are burned. They include a number of heavy metals (such as mercury, cadmium and lead) which are extremely toxic to humans. This means that each year around 53 tonnes of mercury are released into the environment as a result of coal combustion. Domestic sector energy consumption accounts for about 18 tonnes of these emissions. These emissions will not all reach the atmosphere. Some will be removed in bottom ash and flyash, or in scrubbing systems. It is difficult, however, to recover metal from the ash, or from filtering systems, and the disposal of contaminated ash also represents an environmental problem.

The environmental benefits of energy efficiency

The environmental benefits of energy efficiency are in the form of *avoided environmental costs*. That is to say, by saving energy through energy efficiency measures, it is possible to avoid the environmental impacts which would otherwise have been caused by generating, supplying and distributing that energy.

For many of the environmental problems caused by energy consumption, the time-scale for action is increasingly short. The EC directive on acid emissions from large combustion plant requires reductions in acid emissions by 1993, 1998 and 2003. The UK government has pledged to stabilise CO_2 emissions by the year 2005. International calls for reductions in CO_2 emissions by the same date, may force that time-scale even further. The phasing out of disposal of industrial wastes at sea, and the increasing shortage of landfill sites, will force the issue of fly- and bottom ash disposal. Increased public concern over disposal of radioactive materials in the environment has already blocked the two cheapest disposal routes for nuclear wastes, namely sea disposal and shallow burial.

It is true that there are some technological answers to some of the environmental problems detailed in the preceding sections. On the other hand, these kinds of technological answers have two distinct drawbacks. Firstly, the addition of end-of-pipe measures always has an associated cost. Secondly, these responses often produce their own environmental problems. Rather than reducing environmental burdens, add-on technologies often simply replace one environmental problem with another.

The greenhouse effect provides an example of this. The technological responses to global warming are to seek end-of-pipe decarbonisation or to promote nuclear power, which has no *direct* emissions of carbon dioxide (12). However, the practical possibilities for flue gas decarbonisation are severely limited and extremely expensive. Nuclear power suffers of course from the environmental impacts outlined above. In addition, the cost burden associated with this response is heavy.

Energy efficiency, on the other hand, reduces environmental burdens often at no net cost at all. For the case of carbon dioxide emissions for example, a cost comparison of various demand side energy efficiency options, and various supply side options carried out for Friends of the Earth reveals (Figure 10) that a least-cost strategy for the stationary sectors (ie excluding transport) could reduce carbon dioxide emissions considerably by 2005 without increasing nuclear power commitments, and, in addition, could reduce emissions significantly at a negative net cost to society (13). There would also be environmental benefits of course in terms of acid emissions, particulate matter, heavy metal emissions, land-use, water use, and the environmental impacts of mining, transportation, and distribution.

What then are the specific ways in which energy efficiency can be implemented?
Although it is beyond the scope of this paper to describe the technical parameters of energy efficiency technologies in detail (14), it is worth mentioning the broad areas in which the efficient use of energy can be promoted. As noted above, energy efficiency has both demand and supply side implications. Indeed of the seventeen different options shown in Figure 10 for reducing CO_2 emissions, thirteen are energy efficiency options in the broad sense of the term. Nine of these refer to demand side efficiency measures and the rest to supply side measures.

Supply Side Energy Efficiency
The supply side energy efficiency measures are essentially those measures which supply the same final demand for energy to consumers, but reduce the losses involved in supplying that energy.

The biggest scope for such measures is provided by *combined heat and power* (CHP) generation. For every unit of delivered electricity generated by conventional thermal combustion plants, approximately three units of primary energy are consumed. In other words, two thirds of the primary fuel input is lost as waste heat. In some ways, it is wrong

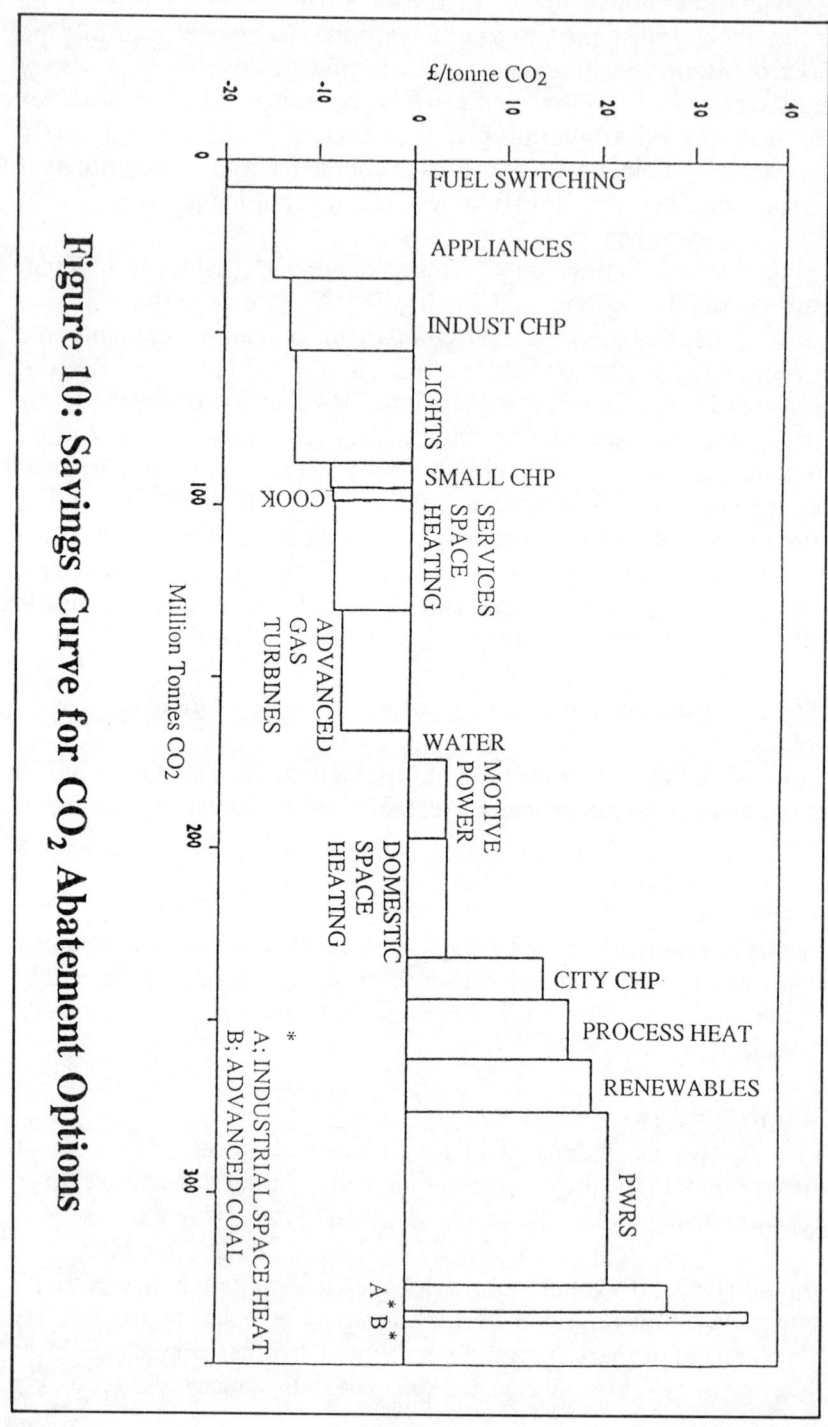

Figure 10: Savings Curve for CO₂ Abatement Options

to classify this as an inefficiency of the electricity generation system. The current efficiency of generation of electricity could be improved from about 35 per cent to around 40 per cent (using advanced coal technology) or 50 per cent (using integrated gas-fired combined cycle turbines), but there are limits imposed by the laws of thermodynamics which reduce the scope for efficiency improvements beyond about 50 per cent, if efficiency is construed in terms of electricity generation. In other words, electricity generation itself is operating more or less at the theoretical limits of efficiency. On the other hand, the energy losses, theoretically inevitable in terms of electricity generation, represent avoidable inefficiencies in terms of overall energy supply, because some at least of the energy loss is recoverable as heat. Combined heat and power generation is the name given to generation plant in which the heat losses incurred from electricity generation are used to supply heat to industry or to domestic consumers for process heating requirements or for space heating.

Combined heat and power (sometimes called co-generation) represents a significant improvement in the efficiency of supply of energy over conventional energy supply systems. The benefits in terms of environmental impact are considerable (see Figure 10). The scope for combined heat and power is great. On the basis of the existing heat demand in domestic, industrial and commercial premises the Energy Technology Support Unit (15) has estimated an 'achievable contribution' amounting to 10 per cent of current electricity demand, although the estimated technical potential was much higher than this.

Demand Side Energy Efficiency
Demand-side energy efficiency measures focus on the way in which energy is used by the consumer at the point of demand. For instance, in the domestic sector over half of the delivered energy is used for space heating requirements. The scope for reducing the demand for delivered energy in the domestic sector is therefore dependent on the efficiency with which space heating and the other end-use services can be provided.

For the case of space heating, efficiency depends essentially on the amount of fuel required to provide a certain standard of warmth in the home. There are three main factors which affect efficiency in this sense. Firstly, there is the efficiency with which space heating devices convert fuel into useful heat. For instance, condensing gas boilers are more efficient than conventional oil- or gas-fired boilers which are more efficient than open solid-fuel fires. Secondly, there is the efficiency with which the space heating provided is controlled. For example, careful thermostatic control of temperature according to patterns of use in offices and homes can prevent heat being wasted during times of no occupancy. Thirdly, the efficiency of provision of space heating can be improved by reducing the amount of heat required to produce a given room temperature: ie, by insulating walls, roofs and windows to high standards.

Similar considerations can be outlined for water heating. Efficiency improvements in water heating include improvements in the efficiency of water heating devices, improvements in the control of water heating, and improved insulation of hot water pipes and tanks.

Whilst the best improvements in efficiency of cooking are either from modification of users' habits (eg, keeping lids on saucepans) or from advanced hob technologies, there is considerable scope for improved efficiency electrical lights and appliances. A recent report (16) concludes that 'replacing today's stock of appliances with the best currently available on the UK market would save 21 000 GWh of electricity, nearly a 40 per cent improvement.' It follows, of course, that such an improvement in efficiency would also save substantially on emissions of carbon dioxide, sulphur dioxide, nitrogen oxides, on the generation of nuclear wastes, and so on.

Environmental Benefits: some quantitative estimates
Clearly, the environmental benefits of energy efficiency measures depend to some extent on where those measures are implemented. For example, appliance efficiency measures would save particularly on all those emissions associated with the electricity supply industries, including nuclear waste, carbon dioxide, acid emissions, heavy metal emissions etc. On the other hand improved insulation would show less reduction of the impacts of electricity supply, since a relatively small per centage of space heating is provided by electricity. Generally speaking the environmental improvements are greater for each unit of electricity saved, than they are for unit savings in other delivered fuel, because of the much greater quantities of primary fuel required to provide one unit of delivered electricity.

What is the extent of the environmental benefits of energy efficiency?
The European Commission has proposed a Community Action Plan for improving the efficiency of electricity use (17). The effects of the implementation of such a programme are indicated in Table 3 below.

It is clear from Table 3 that a 20 per cent reduction in electricity consumption would produce a pro-rata improvement only in the

Carbon dioxide emission reductions	40 Million tonnes	7%
Sulphur dioxide emission reductions	540,000 tonnes	14%
Nitrogen oxides emission reductions	150,000 tonnes	6%
Radioactive waste reductions	6,500 cubic metres	20%
Reductions in mercury emissions	8 tonnes	15%

Table 3: Environmental Benefits of 20% Reduction in Electricity Consumption

reduction of nuclear wastes. Reductions in carbon dioxide and acid emissions are dependent to some extent on efficiency improvements being made with respect to all of the delivered fuels. When the Energy Efficiency Office was first established Peter Walker, then Secretary of State for Energy, informed the House of Commons that the success of the office 'would be judged by the progress it makes in promoting energy efficiency towards a 20 per cent saving by the 1990s'. The environmental benefits of achieving that 20 per cent reduction in energy consumption, are illustrated in Table 4. In this

Carbon dioxide emission reductions	110 Million tonnes
Sulphur dioxide emission reductions	740,000 tonnes
Nitrogen oxides emission reductions	440,000 tonnes
Radioactive waste reductions	6,500 cubic metres
Reductions in mercury emissions	10 tonnes

Table 4: Environmental Benefits of 20% Reduction in Energy Consumption

case, of course, pro rata reductions are made in all of the environmental pollutants.

These estimated reductions are meaningful perhaps only in the context either of detailed scientific knowledge about environmental effects, or else of specific reduction targets for individual pollutants. It is difficult to estimate, for example, what might be the environmental benefit associated with a 740,000 tonne reduction in sulphur dioxide emissions, although some authors (18) have attempted this sort of cost-benefit analysis.

On the other hand, there is certainly some scope for assessing the cost-effectiveness of achieving specific reductions through energy efficiency as opposed to add-on technology. The cost of achieving reductions in sulphur emissions through abatement technology would be considerable (19). On the other hand, energy efficiency options could achieve this at a net negative cost to society. This form of analysis - which we might call cost-effectiveness analysis - is the same kind used to produce the cost curve shown in Figure 10. Energy efficiency is not always the only way to reduce environmental burdens, but it is invariably the most cost-effective way of achieving specific reduction targets.

Environmental Impacts of Energy Efficiency
Finally, a few words should be said about the environmental impacts of energy efficiency. Net savings in energy consumption will generally mean that environmental burdens are lessened by implementing energy efficiency measures. However there are some concerns over specific aspects of energy efficiency technologies which should not be neglected. Some insulation materials, for example, can create health hazards. Equally, there is at present some trade-off between CFC use and efficiency of refrigerators (20). Another example is provided by energy efficient lighting. Compact fluorescent light bulbs can save up to 75 per cent of the electricity required for incandescent bulbs (21). But they also require the use of mercury. An analysis carried out at Lund University in Sweden (22) suggests that the use of compact fluorescents results in net savings in mercury emissions despite the use of mercury in the fluorescent. Nevertheless, it is essential that energy efficiency measures are constantly vetted for potential environmental impacts. Otherwise, there is the risk of simply replacing one kind of environmental problem with another.

Implementing energy efficiency

A recent report from the House of Commons Energy Committee (23) has endorsed Peter Walker's 20 per cent figure as a national target for improved energy efficiency. The committee noted however that there was 'no room.. for any complacency about the rate of improvement in the country's energy efficiency and the likelihood of achieving the present national target.'

The report also discusses in some detail the barriers to the implementation of energy efficiency and the measures which it believes the government should consider in implementing an energy efficiency policy. It stresses particularly the need to take account of the environmental benefits of energy efficiency, 'especially reduced CO_2 emissions, which cannot be measured in terms of energy expenditure saved'.

Given the obvious economic benefits of energy efficiency, it is at first curious that the potential for energy efficiency is not already exhausted. Closer examination reveals, however, that there are a number of different barriers which are currently operating as obstacles to the implementation of energy efficiency (24).

Obstacles

First there are what might be called 'problems of market structure'. These problems include the imbalance between the market for energy supply and the market for energy efficiency, the price regulations for the energy supply utilities, and the separation of benefits from costs (the so-called 'tenant/landlord problem').

Second, we can identify problems associated with the lack of information about, awareness of, and technical expertise in energy efficiency. Third, there is the problem of what might be called 'irrational' behaviour on the part of consumers: the failure of householders and firms to maximise profit in economic decision-making processes even when economic advantages are proven and when information and expertise are available.

The Structure of the Energy Market

Historically, the energy market has grown up around the energy supply companies. Marketing energy supply is now relatively straightforward. Well-established systems exist for marketing electricity, gas and oil. These systems rely on relatively simple market structures often involving centralised technology bases and centralised or regional utilities and decision-making structures. Transactions between consumer and supplier are therefore relatively straightforward, and quantities and values of the commodities being bought and sold are generally understood by both producer and consumer.

By contrast, the market for energy efficiency is historically under-developed and inherently more complex. Implementing energy efficiency depends crucially on the choices of a large number of widely disaggregated consumers, all with differing profitability requirements and preferences, and each of whom is continually confronted with a variety of decisions relating to energy-using activities.

This situation has two specific impacts. First, it means that in most countries there is at best a haphazard or severely limited market for energy efficiency services. Second, investment in energy efficiency is undertaken, if at all, by households and firms at discount rates considerably higher than those used by the utilities when they invest in energy supply.

Clearly, one way of overcoming this imbalance would be for energy efficiency investment to be undertaken by energy utilities at the same rate of return as energy supply. This would allow energy conservation to be placed on the same footing, as it were, as energy generation. But in the UK this solution is impossible because of the price regulation regime of the supply companies.

Price Regulation of Utilities

Historically, energy utilities have made money from selling energy. Profitability requirements (whether for a nationalised industry or for a privatised industry) dictate that energy sales increase rather than decrease. Privatisation is likely to worsen this effect because profitability requirements are the prime motivation for a private enterprise, whereas a national government could, if it so chose, legitimately reduce its own profitability requirements explicitly to serve social or environmental goals.

Price regulation, rather than mitigating this problem has tended to exacerbate it. In the UK for example, electricity cost increases which a Regional Electricity Company is allowed to pass on to its customers are regulated according to a formula:

$$RPI - x + y$$

where RPI is the retail price index (a measure of inflation), x relates to efficiency improvements and y relates to costs which may be passed directly onto the consumer or which are separately regulated.

Efficiency here of course relates not to energy efficiency but to the administrative efficiency of suppliers in operating the service. The problem with this formula is that it leaves no margin for recovering the capital costs of energy efficiency from the consumer. If investment in energy efficiency takes place, then generally speaking the total amount of energy distributed will fall, but the overall costs of that distribution will not fall as a proportion, because many of the associated distribution costs are fixed. This means that to recover all of the distribution costs a higher unit cost for distribution will be necessary. Price regulation does not, as it stands, permit this price increase.

The 'tenant/landlord problem'

The third important structural impediment to the uptake of energy efficiency is the so-called 'tenant/landlord problem'. The problem can be stated, generically, as the problem of separation of responsibility for investment expenditure from operating costs and benefits. It gets its name from the typical situation in which it occurs: a landlord is responsible for investment decisions concerning implementation of energy efficiency measures (such as insulation or double glazing), but it is the tenant who is responsible for the fuel costs. Thus, the tenant stands to benefit from the installation of energy efficiency through lower fuel costs, but is not generally in a position to invest in the required demand side measures. The landlord bears responsibility for such investments, but is unlikely to see a direct return on that investment in terms of energy cost savings.

Another manifestation of the same problem arises in the division between investment and operating costs that occur at the building design and construction stage. Builders, designers and architects are ideally situated to ensure the highest standards of energy efficiency in insulation and design, and to install the most energy efficient appliances and systems. Generally speaking however, they have no direct financial incentive for doing so. Variations of the same phenomenon are evident in many businesses and public sector bodies.

Imperfect Information and Awareness

The efficiency of allocation of resource is of course a cornerstone of the free market philosophy. However, this socially optimal allocation of resources requires that the market operates under conditions of perfect information. This is clearly not the case in the energy market as it now stands. While information on energy supply options must be as near perfect information as it is possible to get in a real market, information and awareness of demand side measures is severely constrained.

Particularly amongst private householders, small and medium-sized companies and small public administrations, there is often either a lack of awareness of the need for energy saving, or else a lack of specific technical information on energy saving technologies. For example, a recent study on attitudes to energy efficiency in the home (25) found that most people 'have little idea that there are major savings to be made through increased efficiency, or routes to it other than through insulation.'

In addition to this general lack of awareness, levels of technical expertise may be insufficient to facilitate the up-take even of known energy saving technologies. Even where information is available on the energy performance of technical goods and services, it is often too theoretically oriented and not accessible to domestic consumers or general management operating under manpower constraints and time pressures (26). As a consequence, energy investment decisions tend to be made without due consideration of the necessary information.

There is currently no energy labelling of appliances in the UK, so that routine purchases are not accompanied by any consideration of energy savings. Television and media energy advertising tends to be dominated by supply company advertising. There is some door to door and mailshot advertising of energy saving devices (double glazing, for example). Consumer reaction to this sort of advertising is cautious however, understandably so, since no comparative basis is available, and such approaches have a reputation for being unscrupulous and over-priced.

Generally speaking, therefore, there is no adequate basis of information either to provide general awareness of the need for or benefits of energy efficiency and conservation, or to advise on the technical and economic details of such measures. The average consumer lacks any technical expertise in these matters and is often in no position to make financial decisions in his or her own best interest.

Limits to Rationality

If firms and households behaved 'rationally', according to the economic theory of the free market, they would attempt to maximise profits or benefits within the framework of their choices concerning the provision of energy services. Yet it is evident that, even where information and expertise are available, this may not always happen.

It is possible to identify a number of reasons for this apparently irrational behaviour. Amongst these reasons we find aspects of the tenant/landlord problem, the fact that energy costs represent a rather small proportion of total running costs and are not therefore subject to detailed economic rationalisation, the fact that in many cases energy supply is paid for by direct billing and remains largely invisible to the consumer, problems of conceptual and institutional inertia, or plain mistrust of unknown technologies, and considerations of preference which are not easily quantified in economic terms, such as status, style and aesthetic choice.

Mechanisms and Instruments

In overcoming the obstacles detailed in the above discussion, there are several different kinds of mechanisms available to policy-makers. First, regulatory measures can improve the structure of the market in energy services. Second, information campaigns can be designed to improve awareness and technical expertise in energy efficiency. Third, economic incentives can be used to prime the market for energy efficient technologies. Fourth, research, development and demonstration schemes can play a crucial role in furthering the technical potential for energy efficiency and in promoting wider awareness. Fifth, government has direct access to public sector buildings, and could 'lead by example' in the implementation of energy efficiency in all public sector buildings.

For the energy sector in particular, there are some specific ways in which regulatory reform and innovation could significantly improve the efficiency of the market with respect to the allocation of resources.

First, energy supply companies could be encouraged to broaden their remit and invest in energy efficiency. Typically this could be achieved by requiring energy utilities to adopt a *least-cost integrated planning approach* to energy services, in which the utility must adopt the cheapest method of providing for a marginal increase in energy demand whether by energy conservation or energy generation. Similar approaches are already operating in some regional utilities in North America (27). A number of ways of achieving least-cost integrated planning through the formation of independent 'third-party' energy service companies have also been proposed (28). These methods would subject energy efficiency investments to the same rate of return as energy supply, so overcoming the first of the structural obstacles described above.

Second, it is clear that the *price regulations* governing utilities should be changed to allow energy supply companies to benefit financially from undertaking energy efficiency investments. Two possible approaches can be taken to this. The first approach, which has been proposed by the Public Utilities Commission in Maine, is to regulate energy utilities by controlling the allowed rate of return according to how well or how badly the companies behave in terms of saving the consumer money. Utilities with a good record in this respect would be allowed a higher rate of return on lower electricity sales; those with a poor performance would be allowed lower profits. This sort of regulation would allow utilities to invest in energy efficiency and recover the cost of their investment through consumer bills. Another approach would be to alter the existing price regulation formula, in such a way as to allow for and encourage investments in energy efficiency. Such an approach has been proposed for the gas supply industry in the UK for example, where price regulation of the form:

$$\text{RPI} - x + y + e$$

will come into effect from April 1992. Here *e* relates directly to energy efficiency expenditure.

Third, *building and housing regulations* could force landlords and builders to apply high energy efficiency standards to overcome the tenant/landlord problem. From the point of view of cost-effectiveness, it is clearly preferable to reduce energy consumption (and hence carbon emissions) by energy efficient design and construction than by retrofitting insulation measures. This is not to suggest that retrofitting of insulation measures is not important. In view of the slow rate of turnover of stocks, it is clearly essential. Nevertheless, costs are significantly lowered if energy efficiency measures are incorporated at the design stage. Furthermore, the introduction of building regulations would lead to improvements in economies of scale in the market, and reduce the transaction costs associated with demand side management by reducing the number of actors involved.

Fourth, energy-using products could be required by law to be labelled with information on energy consumption and running costs (*energy labelling*). In the US this is already a requirement for domestic appliances; Denmark requires similar information by those selling or letting houses and offices. Such labelling could accompany information campaigns, advertising, on the spot consultations and advice, and systematic education and training programmes to improve the flow of information in the energy services market.

Fifth, a legal obligation could be imposed on companies over a certain size to conduct an *'energy audit'* to identify potential savings. (For households, such audits could be provided as a free service by local environmental conservation agencies.)

Sixth, energy efficiency standards could be applied by law to appliances (*appliance efficiency standards*), whereby standards are set for the performance efficiency of energy appliances, and manufacturers are required to meet these standards before offering their appliances for sale. Similarly, energy efficiency standards can be required of new industrial equipment and processes. In their final report, the Select Committee on the European Communities specifically recommends that 'satisfactory standards and testing methods for domestic appliances should be worked out, and, when the time is right, imposed on Community manufacturers by legislation' (29), and particularly notes the lead taken in the United States by the National Appliance Energy Conservation Act (1987) which sets 'base minimum [standards] below which various electricity consuming items shall not fall'.

In addition to these regulatory reforms, there is also scope for the use of financial incentives - both positive and negative. Arguments have been made for the use of a *carbon tax*, that is a tax on fossil fuels, specifically related to the carbon content of those primary fuels. This tax is seen as one way of internalising the external costs of environmental damage from carbon dioxide emissions. Similar arguments have been made for acid emission taxes.

Several points are worth making about environmental taxation. In theory, the idea of internalising external environmental costs is well-founded. The environment is what is called by economists a common property resource - in other words, a free good, which can be used by everyone. Since people can use this resource freely, they tend not to optimise their investments according to the environmental damages which they incur. Since environmental damage does not belong to the project costs, there is no incentive to invest in ways to reduce that damage. By charging for use of the environment (ie through a carbon tax or an acid emissions charge), so the argument goes, environmental costs will be internalised and an optimal allocation of environmental resources can be achieved.

The practical difficulties of this approach are well illustrated by the example of the carbon tax. In theory, the tax should be set at a level which reflects the environmental damage caused by carbon emissions. However, it is impossible in practice to calculate what those costs might be. Moreover, even if we could calculate such costs, there is no guarantee that the appropriate level of taxation would be sufficient to achieve the required reductions in carbon dioxide. This is because the price elasticity of energy demand is very low. In other words, energy prices must go up a long way before energy demand falls. The reasons for this are exactly the same as the reasons outlined above for the failure of the market to implement cost-effective energy efficiency measures. The market is distorted in favour of energy supply and against energy efficiency. Economists have therefore argued in favour of taxation levels that are determined not on the basis of environmental damage costs, but rather on the basis of achieving specific reduction targets. The level of taxation is then determined in part by the required reduction and in part by the elasticity of the market.

The problem with this latter approach is that it results in extremely high taxation rates. For instance tax rates of over 300 per cent of the cost of coal have been calculated in order to reduce carbon emissions by 20 per cent before the year 2005 (30). These kinds of taxation rates are likely to prove both politically unacceptable, and to have a higher impact on low-income households than on well-off ones. Moreover, there is no justification for them, either in terms of the environmental costs of damage or in terms of the efficiency of allocation of resources. If the market is distorted, then resources will not be efficiently allocated, even in the presence of high taxation rates. Indeed, the imposition of high taxation rates in a distorted market could result in a less efficient allocation of resources, rather than a more efficient one.

This is not to suggest that there is no role for environmental taxes. On the contrary, environmental taxes may indeed be a useful way to influence the behaviour of the market. They may also be an essential mechanism for raising revenue which can then be used to finance information programmes, regulatory reform, R&D and financial assistance schemes. These schemes include grants and subsidies for home insulation such as those provided by the government-funded Home Energy Efficiency Scheme (HEES), which came into effect in January 1991. The lesson to be learnt here, however, is that the first thing to do is to remove the obstacles currently preventing the implementation of cost-effective energy efficiency in the market. This means designing and implementing the kinds of structural and regulatory reforms outlined above. Until these kinds of reforms have been carried out there will be little hope of realising the potential for cost effective energy efficiency improvements, and therefore little hope of securing the enormous environmental benefits which energy efficiency could provide.

References

1. The analysis in the section is based on data from: Department of Energy, *Digest of UK Energy Statistics 1990,* HMSO, 1990; Department of Energy, *Energy Use and Energy Efficiency in the UK Domestic Sector up to the year 2010,* EEO, Energy Efficiency Series II, HMSO, 1990; House of Lords Select Committee on the European Communities, *Efficiency of Electricity Use,* HMSO, 1989.

2. See in particular the forecasts in the EEO series: *Use and Energy Efficiency in the UK Manufacturing Industry up to the Year 2000,* vol 3 in the Energy Efficiency Series, Energy Efficiency Office, Dept of Energy, 1984; *Prospects for the use of Advanced Coal Based Power Generation Plant in the UK,* Energy Paper 56, Dept of Energy, HMSO, 1988; *Energy Use and Energy Efficiency in the UK Commercial and Public Buildings up to the Year 2000,* No 6 in the Energy Efficiency Series, Energy Efficiency Office, Dept of Energy, 1988; *Energy Use and Energy Efficiency in the UK Domestic Sector up to the Year 2010,* Energy Efficiency Office, Energy Efficiency Series 11, HMSO, London, 1990.

3. Jackson, T. and Roberts, S. *Getting out of the Greenhouse,* Friends of the Earth, 1989.

4. Elsworth, S. *Acid Rain,* Pluto Press, 1984. Owens, S. and Owens, P. *Environment, Resources and Conservation,* Cambridge University Press, 1991.

5. CEC, *Large Combustion Plant Directive,* Commission of the European Communities COM (88) Official Journal L336 vol 31, Nov 1988.

6. House of Lords Select Committee on Science and Technology, *Greenhouse Effect,* HL paper 88, House of Lords Select Committee on Science and Technology, Session 1988-89 6th Report, 1989.

7. Mitchell, C. 1991 'Coal-bed Methane in the UK', *Energy Policy,* 19(9), Nov. 1991, pp. 849-854.

8. Mitchell, C. Sweet, J. and Jackson, T. 1990, 'A Study of Gas Leakage from the UK Natural Gas Distribution System', *Energy Policy* vol.18(9),1990.

9. Jackson, T. 'Renewable Energy: Great Hope or False Promise?' *Energy Policy* vol 19(1), Jan/Feb 1991.

10. IAEA, *Assessment of the Radiological Consequences and Evaluation of Protective Measures,* The International Chernobyl Project, International Atomic Energy Agency, Vienna, May 1991.

11. Friends of the Earth, *Nuclear PR Agency Helped to Write Official Chernobyl Health Report*, FoE Press Release, 23 May 1991.
12. Mortimer, N. 'Nuclear Power and Global Warming' *Energy Policy* vol 19(1),1991.
13. Jackson, T. *The Role of Nuclear Power in Global Warming Abatement Strategies,* Evidence to the Hinkley Point Inquiry; Friends of the Earth 1989. Jackson, T. 'Least Cost Greenhouse Planning',*Energy Policy* vol 19(1),Jan/Feb 1991.
14. For more details see Department of Energy, *Energy Use and Energy Efficiency in the UK Domestic Sector up to the year 2010,* HMSO, 1990; Jackson, T. and Roberts, S., 1989, op.cit. ref 3.
15. Department of Energy, *Environmental and Economic Implications of Small-scale CHP*, Energy and Environment Paper 3, R Evans, ETSU, Dept of Energy, March 1990.
16. Department of Energy, *Energy Efficiency in Domestic Sector Electric Appliances,* Energy Efficiency Office, Energy Efficiency Series 13, HMSO, London, 1990.
17. See House of Lords select Committee on the European Communities, op.cit. ref.1
18. Hohmeyer, O., *Social Costs of Energy Consumption,* Springer Verlag, 1988.
19. See Owens and Owens, 1991 op.cit. ref 4.
20. See Department of Energy, 1990, op.cit. ref.14.
21. There has been some dispute over the extent to which compact fluorescents generate real savings in generating capacity and therefore real financial savings to generators. The reason for this dispute is that compact fluorescents make a demand for 'reactive' power if they are not corrected for power factor. In general however factor correction is possible, and in any case the demand for primary fuel consumption depends on 'active' rather than reactive power demand. In these terms the advantages of energy efficient light-bulbs are real.
22. Mills, E. 'Evaluation of European Lighting Programmes',*Energy Policy,* vol 19(3), April 1991.
23. House of Commons Energy Committee, *Energy Efficiency,* 3rd Report, Session 1990-1991, HMSO, London 1991.
24. Jackson, T. *Efficiency without Tears: 'no-regrets' energy policy for a warming world,* Friends of the Earth, June 1992; Jackson, T. and Jacobs, M. *Carbon Taxes and the Assumptions of Environmental Economics,* in T. Barker (ed), *Green Futures for Economic Growth* T Barker (ed), Cambridge Econometrics, 1991.

25. A. Hedges, *Attitudes to Energy Conservation in the Home,* Department of the Environment, HMSO, London, 1991.
26. Jochem, E. and Gruber, E. Obstacles to Rational Electricity Use and Measures to Alleviate Them, *Energy Policy,* vol 18, No 4, May 1990.
27. ASE, *Designing and Evaluating Demand Side Management Rebate Programs,* Alliance to Save Energy, Washington, USA, 1988
28. R. Williams, 'Innovative Approaches to Marketing Electric Efficiency', in T. Johansson et al (eds.), *Electricity: Efficient End-Use and New Generation Technologies and their Planning Implications,* Lund University Press, 1989.
29. House of Lords Select Committee on the European Communities, op.cit. ref.1.
30. See T. Barker(ed.), 1991, op.cit. ref.24.

Summary of seminar on environmental aspects of energy efficiency

Chaired by: Sir Terence Heiser, Permanent Secretary, Department of the Environment.

Speaker: David Gee, former Director of Friends of the Earth; independent environmental consultant.

The seminar was addressed by David Gee, then Director of Friends of the Earth, who began by observing that improvements in energy efficiency were 'unique in policy terms' in that they could increase social welfare, improve economic efficiency and employment prospects, and help to avoid environmental disasters potentially generated by global warming. 'Racking my brain failed to produce another policy goal that achieves the same sweeping range of fundamentally "good" outcomes'.

The focus of the talk was on the barriers to effective action to realise the many environmental and other benefits of energy efficiency, and on ways of overcoming the obstacles. David Gee reviewed the environmental costs of intensive use of fossil fuels and noted the 'overriding environmental imperative' for industrialised countries to reduce energy consumption. He noted that this is often thought of as a cut in energy demand, but this was misleading. No-one actually wants energy itself – they want to consume services provided by energy supplies, such as warmth, hot water, cooked food. Improving the energy efficiency of the systems and appliances delivering these services is a means of reducing fuel consumption while maintaining current levels of energy services.

The case for making major improvements in energy efficiency was justified on economic grounds even without taking into account the costs associated with environmental pollution and carbon dioxide generation. However, 'like no other insurance policy, in the case of improving energy efficiency, the national economy as a whole actually saves money by insuring against the risk of environmental damage'. He paraphrased the energy researcher Amory Lovins of the Rocky Mountain Institute: 'Improving energy efficiency is not just a free lunch, it's a lunch you are paid to eat'.

Given the evident benefits, what was constraining greater activity in this area? Technical and macroeconomic considerations did not block progress, but political obstacles did. David Gee saw these primarily as failures by politicians 'to understand the true nature of energy efficiency as an option in energy policy' and to develop a vision of how to deploy a comprehensive range of measures to realise the potential benefits.

The failure of understanding related to the lack of appreciation of 'the equivalence of improved energy efficiency with displaced energy supply'. Substituting a 20 watt compact fluorescent light bulb for a 100 watt incandescent one not only produces the same amount of light, but cuts the amount of power needed at the power station: the compact bulb runs on one-fifth the amount of electricity and thus reduces associated power needs and pollution by 80. 'If this equivalence holds, then the political question becomes: why not establish a market for energy services where energy supply and energy saving compete... on equal terms?'

The failure of vision, David Gee argued, was related to the lack of awareness of the need for a wide-ranging set of measures to overcome the numerous obstacles to change. He set out the key barriers to greater energy efficiency in the UK:

– economic obstacles, including many householders' lack of capital to pay for investments such as cavity wall insulation, and the unbalanced imposition of VAT on energy efficiency products but not on fuels;
– gaps in information and expertise, notably the lack of standard energy labelling of appliances and buildings;
– lack of ready availablity of energy efficient technologies to consumers;
– structural obstacles, notably the incentive for gas and electricity companies to encourage sales of more fuel rather than investment in energy efficiency and conservation;
– psychological and social factors, including consumers' lack of understanding of the links between energy consumption and environmental damage and lack of motivation to invest in energy efficiency measures.

The implications of this multi-faceted set of problems were 'somewhat uncomfortable' for policy makers, 'because it points to the need for a multifactorial approach to policy, tailoring certain measures to encourage the uptake of particular technologies to specific sectors. The easy slogans evaporate'. The different barriers to change could be overcome through a mixture of grants, fiscal incentives, regulatory changes, minimum efficiency standards, information and training campaigns, institutional restructuring and direct government action. However, this tailored programme of measures would call for 'an uncommon breadth of political vision and ideological humility', and would need to be devised within the framework of a national target for the reduction of harmful emissions. A national target would offer a clear signal to commerce and industry and act as a spur to policy makers in all areas.

Discussion

The discussion centred on the limitations of the current energy market in relation to maximising the scope for energy efficiency and consequent environmental benefits. Speakers noted numerous constraints on the development of an energy-efficient economy:

– Reservations about the economic impact of 'a go it alone' policy - either at national or European level - and the limited success of such a policy in the context of the rising international trend in energy related CO_2 production.

– How far can economic policy instruments alter consumer choices? It was argued by one speaker that consumers do not invariably act on a 'least-cost' basis in relation to energy consumption, and it would be naive to assume that economic measures alone could produce the desired changes in behaviour in the direction of conservation and greater household investment in insulation.

– The inadequacies of information about energy efficiency and conservation were agreed by many speakers to be a key instance of market failure in the energy sector. It was noted that energy efficiency advice was not readily available to the consumer from sources such as retailers and companies; and in addition, manufacturers did not provide standardised information on the energy consumption levels of their appliances. Some speakers avocated that the generating companies should be building up a market in 'energy services' rather than simply in energy supply.

– Some speakers argued that the failure of energy efficiency to 'take off' reflected a basic clash between those wedded to an ideological view of the superiority of free markets and the demands imposed by energy infrastructure and environmental challenges. Taking least-cost planning seriously as a means of optimising use of energy resources implied a more centralised, interventionist approach to managing energy supply and demand than had been acceptable to the Conservative government after 1979 not withstanding the significant subsidy given to the nuclear industry.

The discussion also focussed on the degree to which environmentalists were justified in claiming that energy conservation and efficiency measures represented a relatively painless approach to achieving environmentally sustainable development. One speaker argued that the policies they proposed would have a substantial effect in terms of the diversion of resources. Economists and environmentalists were 'still talking across one another'. An optimistic analysis produced for Friends of the Earth of costs and benefits from energy efficiency investment (1) failed to take into account 'hidden' transition costs that would be incurred by companies, households and government in the process of switching fuels and implementing a national energy efficiency programme. Another speaker took up this theme and argued that one reason for the lack of enthusiasm for major energy efficiency investment as a response to fears over global warming was the uncertainty over the benefits that would accrue as a result. The large costs of any programme of investment were clear, but the issue was still clouded by uncertainty and the scale of the economic benefits was not evident to policy makers.

In response, Tim Jackson, author of the seminar paper, agreed that some transitional costs were hidden in the analysis for Friends of the Earth, notably the 'profitability costs' incurred by companies needing to make investments in energy efficiency and fuel switching. He said that there was a major question over the discount rates that need to be applied to investments in energy infrastructure. In general advanced industrial countries had used discount rates to build up a system that spreads costs over long periods and delivers energy on a very large scale. He argued that institutionally we have never had to consider environmental impacts from energy generation and thus integrate them into discount rate calculations: this 'institutional infancy' had to be overcome if progress was to be made in achieving the benefits of energy efficiency and conservation.

David Gee concluded by saying that a dialogue between environmentalists and economists had now begun in earnest in order to clarify these issues, but emphasised that there was no escaping the need to go beyond 'marginal' policy changes and face up to the challenge of fundamental rethinking of energy policy in the light of global environmental dangers.

Reference

(1) T. Jackson, 'Least Cost Freenhouse Planning', *Energy Policy*, vol.19(1), Jan./Feb. 1991; see also T. Jackson, *Efficiency without Tears: no regrets energy policy for a warming world*, Friends of the Earth, 1991.

Economic aspects of energy efficiency

John Chesshire
Head, Energy Policy Programme
Science Policy Research Unit
University of Sussex

Introduction
This paper addresses in a largely non-technical manner some of the key economic issues of relevance to energy efficiency policy in the UK, particularly with respect to energy consumption in the domestic (residential) sector. The paper begins with a range of definitions essential for understanding the economic dimensions of the subject, and with an overview of recent trends in energy use in the domestic sector. This is followed by a brief analysis of the main mechanisms and actors involved in improving energy efficiency in this sector. It then examines justifications for an energy efficiency policy, as well as the Government's current, essentially laissez-faire and market-oriented, policy stance. The paper concludes by identifying a number of issues of relevance to energy efficiency policy over the coming decade.

Some key concepts and definitions
The demand for energy is a derived one. No one buys fuel or electricity for its own sake (even speculative traders in futures markets). Instead, energy is purchased as a means of undertaking productive processes and of obtaining movement, communication, heating and cooling, lighting and cooking. In other words, it is the services made possible by the direct or indirect use of energy which are sought by consumers.

This important characteristic of energy demand leads to an immediate institutional difficulty in approaching policy formulation for energy efficiency. Every household and company, every motorist and office worker is involved in energy use and/or purchase decisions. Likewise the responsibility for the formulation and execution of energy efficiency policy is inherently fragmented, because all government departments are either directly involved in activities which consume energy or - indirectly, via their respective responsibilities - have oversight of energy suppliers, users and equipment manufacturers.

To avoid subsequent confusion in debate, it is necessary to define some key concepts which shape the economic dimension of energy efficiency policy, whether for the domestic or any other demand sector.

Energy efficiency relates to the reduction of energy inputs per unit of output - whether the latter is measured in terms of useful heat, light and mechanical work, miles per gallon or industrial production. It is important to distinguish between the technical and economic potential for energy efficiency. Estimates of the technical potential indicate how much energy could be saved by the widespread application of current best-practice or readily identified future best-practice technology. On the other hand, the economic potential indicates how much could be saved within specified investment criteria (such as a rate of return or payback period). As a result, the economic or cost-effective potential is dependent upon judgements about current and expected fuel prices, the availability and cost of capital, technological change and the underlying rate of replacement of capital stocks of energy using appliances, plant and equipment.

In principle, the proportion of the technical potential for energy efficiency which is likely to be economically attractive will increase over time in response to changing relative factor prices - ie energy, capital (both as the cost of money and that of capital equipment), labour and raw materials. For an economist, investment criteria are obviously significant. However, there is a danger of over-emphasising their importance. For example, it is not clear how consumers evaluate paybacks for fitted kitchens or new bathroom suites - yet these involve financial commitments comparable to major energy efficiency improvements in the home.

A number of other useful concepts relate to the economics of energy efficiency. *Inter-fuel substitution* (fuel switching) involves the reduction in use of an expensive energy resource and its substitution by a cheaper, usually more plentiful, resource. Thus it is possible, through fuel switching, to save money without any corresponding reduction in the thermal quantity of fuel used.

Depletion policy aims to influence the rate of exploitation of a natural resource in actual or potential short supply (1). If considered desirable, a depletion policy can act on both the supply and the demand (energy efficiency) sides to curb rates of extraction and use of a scarce commodity.

Some commentators in the energy efficiency debate have chosen to imbue the drive for energy efficiency (or, as they might prefer to call it, *energy conservation*) with a more fundamental moral dimension. Mindful of the finite nature of fossil fuel reserves, or of the limited capacity of the universe to diffuse or absorb energy-related pollution, these commentators have powerfully advocated the reduction in energy use at almost any cost. Without doubt, such an approach raises many challenging questions, including those relating to inter-generational equity and the extent to which the economist's conventional set of "value-free" analytical tools (such as social time preference and the use of appropriate discount rates) can handle these essentially moral and philosophical questions.

Trends in UK domestic sector energy demand

Energy demand in the UK domestic (residential) sector has grown only slowly over the past 20 years. Given the dominance of space heating in total demand, there have been marked peaks in consumption in cold winters. Demand growth in the 1970s was associated with increased market penetration of central heating and rising comfort levels (whole house temperatures). For example, the proportion of households with some form of central heating, mainly gas-fired, rose from 38 per cent in 1972 to nearly 70 per cent in 1985 and has continued to increase since. In addition, as a result of population growth and in particular of demographic changes which have affected household formation, the number of households has expanded - an underlying factor which is expected to continue at least until the end of the century.

Energy demand growth in this sector has been influenced, especially in the 1980s, by the slow but increasingly widespread adoption of a range of energy efficiency measures (eg draught stripping, loft and hot water tank insulation, cavity wall fill and double glazing). More stringent Building Regulations have led to higher insulation standards for new buildings, and technological changes (eg lower temperature wash cycles and better insulated cookers and refrigerators) have reduced unit electricity consumption for many electrical household appliances. Recent trends in domestic sector energy consumption are summarised in Table 1.

TABLE 1. Trends in UK Domestic Sector Energy Demand

(billion therms)

	1970	1975	1980	1985	1990
Gas	3.5	5.9	8.4	9.7	10.2
Oil	1.3	1.4	1.1	0.9	1.0
Solid fuels	7.1	4.3	3.3	3.1	1.7
Electricity	2.6	3.0	2.9	3.0	3.2
TOTAL	**14.6**	**14.7**	**15.8**	**16.7**	**16.1**
% of total final energy demand (all sectors)	25.3	26.3	28.0	29.6	27.1

Source: Digest of UK Energy Statistics (various years)

The biggest fuel switches over the past two decades have been the decline in coal and the steep rise in natural gas consumption. The replacement of open coal fires, the spread of smokeless fuel zones and consumer preference for cleaner, more convenient and controllable fuels have led to the steady displacement of coal in the domestic market. However, coal use remains of significance in the more remote areas of the UK not served by the natural gas distribution system, by current and retired coal miners in receipt of concessionary coal, and in a small number of collective district heating schemes.

Unlike much of continental Europe and North America, oil consumption in the UK domestic sector has never been large. This was essentially because few households had installed central heating systems prior to the widespread availability of natural gas from about 1970. Oil use in the domestic sector is now concentrated in rural areas, and/or confined to paraffin heaters and heating equipment nearing the end of its working life. Oil's market share is thus expected to continue to decline, primarily as a result of competition from natural gas and electricity.

The development of the off-peak (mainly night-time) tariff in the early 1960s and of night storage heating systems (White Meter and subsequently Economy 7) underpinned the rapid growth of electricity sales over the following decade. However, in the 1970s electrical space heating was in retreat given the rise in fossil fuel prices, delays in commissioning nuclear capacity and the availability of gas. As a result, for example, off-peak electricity sales in England and Wales in

1984/85 were some 35 per cent below their peak year in 1974/75. More recently, active marketing campaigns have reversed this decline in electricity sales for space heating, especially in new all-electric dwellings. Increased appliance ownership, particularly of higher load appliances such as washing machines, fridge-freezers and dishwashers, has also caused a resumption of growth in electricity sales.

Natural gas has been the major beneficiary of fuel switching in the domestic sector over the past two decades. This is highlighted by the rise in its market share, from 24 per cent in 1970 to 63 per cent in 1990. At present some 75 per cent of households in Great Britain use natural gas (and, in those areas served by gas, it accounts for some 85 per cent of the available market for central heating). Only about half of the remaining households are located in areas close to the gas distribution network; in particular, natural gas is not available in Northern Ireland and the gas distribution system is far from extensive in rural Scotland, Wales and the South West region of England.

Natural gas and electricity are widely expected to continue to erode the market shares of coal and oil in the domestic sector. Increased penetration of natural gas will continue in Great Britain, though at a slower rate than in the past. It will be determined largely by the connection policy of British Gas in those areas not presently served by it. Gas is continuing to win the lion's share of new central heating systems, whether installed in new dwellings or retrofitted in existing dwellings. Nevertheless, the electricity industry is confident that it can compete effectively in the space heating market for new dwellings and continue to increase sales for use by those higher load domestic appliances which have yet to achieve high market saturation (eg dishwashers and tumble driers).

However, replacement of heating systems and older less-efficient appliances - combined with further improvements to the building fabric and more widespread diffusion of better instrumentation and controls (eg thermostats) - are expected to offset energy sales growth despite a continued rise in average whole house temperatures over the next decade or two. The average age of the UK housing stock, and its poor thermal performance compared with that in many other countries, suggests that much scope remains for further cost-effective energy efficiency investment. The Government itself, in evidence to the Intergovernmental Panel on Climate Change (IPCC), stated that:

> a typical householder can achieve a 20 per cent saving in energy
> by investments with what he or she may regard as an acceptable
> payback period (3 to 5 years at most)...On the payback criteria used

by the Government for its own investments, the percentage saving could rise to 30 per cent and, with longer paybacks, further savings are possible (2).

It is against this context of broad energy use patterns and the considerable scope which still exists for enhanced energy efficiency that we now turn to examine energy efficiency mechanisms, actors and policy developments in the domestic sector.

Some key market mechanisms and actors

The main market mechanism at work in enhancing energy efficiency in the domestic and other sectors is the installation, replacement and retrofitting of energy using structures, appliances and equipment. This underlying economic process of capital stock rotation is crucial but, given the longevity of certain assets (particularly dwellings), its effects on some dimensions of energy efficiency occur over very long time scales. For example, at one end of the spectrum of this timescale is turnover of the housing stock. The energy efficiency performance of new dwellings has improved during the last decade following amendments to the Building Regulations in 1981 and 1989. Even these new Regulations are less demanding than those in force in some other European countries. According to the Housebuilders' Federation, new dwellings built under these Regulations will theoretically save up to 37 per cent of the energy of new houses built before 1981 (3).

However, some 19.7 million dwellings currently in use pre-date these Regulations with (cumulatively) over 3 million built before 1891, 5.9 million before 1918, over 10 million before 1944 and over 17 million built before the first oil shock in 1973/74. No more than about 0.2 million new dwellings are built each year, representing a stock addition/turnover of less than one per cent per year. Thus, whilst yet further improvements to Building Regulations are required, the major route to enhanced energy efficiency in the building stock will always be *retro-fitting of existing dwellings*. Even here, earlier construction standards and techniques may serve greatly to constrain the scope for retro-fitting some measures - for example, only about 18 per cent of the pre-1918 housing stock has cavity walls.

The domestic sector boiler and other primary heating appliance stock is another prime target for energy efficiency measures given the dominance of space heating in total domestic sector energy use. The turnover of domestic gas central heating boilers is about once every 15-20 years but the scope for retro-fitting is negligible. Boiler efficiency can be maintained by regular servicing. Heating *system* efficiency, however, is capable of significant improvement through lagging of hot water tanks and the installation (and correct use) of thermo-stats and other controls. Turnover of many domestic electric appliances appears to lie in the range of 5-10 years. At the other extreme of equipment longevity is the domestic light bulb, with a service life measured in months, and hence capable of much more rapid upgrading to high efficiency types if the economic case is attractive.

The main groups of actors involved include the following:
- a myriad of decisions by *consumers* in (i) their behavioural patterns (lifestyles) in using energy-consuming equipment and appliances and (ii) in their investment decisions as regards new or replacement appliance purchases;
- competition between the *fuel suppliers* for market share on the basis of relative fuel prices, allied - especially in the domestic sector - with cleanliness and convenience. Thus the suppliers' advertising and advice, the range of appliances they sell or advocate, and their leverage upon equipment suppliers to raise thermal or other end use efficiencies play a key role;
- numerous *equipment manufacturers, installers and servicing/repair companies,* in seeking to compete with each other for market share on the basis of price, design, delivery, quality of service, product life, ease of use and energy efficiency;
- a number of key decision makers, particularly *planners, architects and housebuilders,* in determining initial energy efficiency standards for dwellings;
- a number of *regulators,* (including *Offer, Ofgas and the European Commission)* in determining and monitoring energy efficiency standards, services and advice;
- the *Government,* in shaping the framework of fiscal policy (including taxes, grants and allowances) and Building and other Regulations; in the provision of advice and education; and - more generally - for determining strategic policy, setting national priorities, and for measuring and evaluating overall national energy efficiency performance.

The policy background
(a) Some Broad Policy Justifications
Over the past two decades, the broad justifications for formulating an energy efficiency policy have included the following:

- (i) *energy depletion* - the need to conserve fuel given inherent uncertainty about long-term fuel availabilities and prices;
- (ii) *supply security* - the need to minimise the exposure of the UK economy to short-term supply disruptions and/or de-stabilising price hikes;
- (iii) *economic competitiveness* - the need to ensure that use of energy, as with use of capital, raw materials and other inputs, is minimised subject to overall criteria of cost-effectiveness, as a means of boosting the competitiveness of (and of minimising the impacts of energy imports on) the UK economy;
- (iv) *broader social and welfare policy* - to enhance standards of living by minimising unnecessary expenditure by consumers on energy, particularly for those on low incomes;
- (v) *environmental policy* - to assist, together with a wide range of other measures (within and without the energy sector), in curbing the deleterious effects of numerous pollutants upon the local, regional and global environment. Whilst this policy justification is currently the most fashionable, it only serves to strengthen the already powerful underlying case for a coherent, consistently applied, energy efficiency strategy.

Despite these powerful justifications for policy, there is a fundamental inconsistency, particularly although not exclusively in Government, regarding the economic significance of energy efficiency. Developments on the *supply* side of the energy sector relating to coal, oil, natural gas and nuclear power have traditionally been viewed in their fullest *macro*-economic contexts. For example, the policy justifications surrounding decisions on The Plan for Coal (1974), the new nuclear power programme (1979) and the protracted debate and subsequent cancellation of the proposed importation of natural gas from the Norwegian Sleipner field (1984/85) were complex and interwoven. They embraced issues such as:
- import dependence and the strategic vulnerability of the UK economy to disruption in energy supplies;
- the balance of payments and the exchange rate;
- public expenditure and fiscal implications;
- employment and regional policy implications;
- opportunities for furthering technology policy objectives;
- industrial policy questions, including orders for the power plant and off-shore supplies industries, and the need for a shop window at home and other assistance with exports.

Conversely, the strategic and macro-economic justifications for a far more vigorous energy efficiency policy have been almost entirely unrehearsed. Throughout the period since the first oil shock in 1973/74, and notwithstanding the heightened policy attention devoted to this subject in Energy Efficiency Year (1986), energy efficiency has been awarded a far less exalted status. The context for Government energy efficiency initiatives over the past decade has been essentially micro-economic in character - that is, focussed almost exclusively on the family or corporate bottom line, as measured by improvements in household disposable income or company profitability.

Why should there have been this unequal treatment of energy supply and energy efficiency options in the past? The following factors appear to have been of fundamental importance in the UK debate:

– complacency born out of relative plenty (the UK may be described as an island of coal set in a sea of oil) and the UK's effective self-sufficiency in energy, at least over the past decade, alone amongst the major OECD countries;

– the traditional and excessive fixation of the energy institutions and policymakers in the UK with the *supply* side of the energy equation and their comparative lack of experience and competence in addressing *demand* side issues, such as energy efficiency;

– the lack of an appropriate energy policy framework in the UK which has resulted in ad hoc and often short-term decision-making and the observation that, in the recent past, narrow fiscal, PSBR and privatisation preoccupations have overwhelmed virtually all other sectoral policy considerations (4).

(b) Energy efficiency targets

In 1983, when announcing the establishment of the Energy Efficiency Office (EEO), the then Secretary of State for Energy indicated the scope for energy efficiency that he believed could be achieved cost-effectively. The target set was an overall improvement in national energy efficiency of 20 per cent within 10-15 years, equivalent to a reduction in national energy expenditure of £7 billion at 1983 prices. The Secretary of State added that he believed that the UK could go from the bottom to the top of the international energy efficiency league table within the lifetime of the 1983 Parliament (ie by 1988 on a full term). The Secretary of State informed the House of Commons in December 1983 that the success of the EEO "will be judged by the progress that it makes in promoting energy efficiency towards a 20 per cent saving by the 1990s (5)."

In 1983, the EEO provided initial disaggregated estimates, by major end use sector, of the overall 20 per cent national energy efficiency target (6):

Sector	Savings (£ million, 1983 prices)
Domestic	1,900
Public sector	400
Industry/commerce	1,900
Transport	2,800

Following a strategy review in 1985, the EEO calculated that 'normal' market forces might produce savings in energy of £5,040 million per year (in 1983 prices) by 1995, leaving Government programmes to stimulate the balance of £1,960 million per year (7).

In 1989, the National Audit Office (NAO) examined the UK's (and the EEO's) performance against the targets. For the domestic sector, savings achieved by the end of 1987 were estimated at £808 million per year (1983 prices). The NAO judged that this trend, if extrapolated to 1995, could enable the UK to achieve the overall 20 per cent target (8). However, since 1986 real energy prices have fallen, thus reducing market incentives, and environmental protection has become the principal cause for concern.

(c) The government's present policy stance
Many commentators who have been strong advocates over many years of the need for more consistent and serious policy attention to be given to energy efficiency in the UK have experienced what can only be termed a profound sense of deja vu in relation to recent policy discussions about this subject. Those who wish to defend the present Government's 'green' credentials and the strategic significance now accorded to energy efficiency at the highest reaches of Whitehall will point to the fact that, along with renewable energy, energy efficiency warranted half of the brief Annex C of *This Common Inheritance* (9) and that the EEO is now part of the Department of the Environment. However, it must be doubted whether this is a sufficient statement of policy intent.

Against the evident policy need outlined above, it appears that the Government prefers to repeat the now rather stale case for its own continued passivity in this field:

> Energy efficiency measures are by definition in the economic interests of consumers .... The Government therefore starts from the position that consumers do not need subsidy, regulation or financial incentive or penalty (through the taxation system) to get them to act to improve their energy efficiency, as long as the market is working properly (10).

Yet the conflicts inherent in the Government's stance are exemplified by the statement, only a few paragraphs earlier in the same Memorandum, that:

> ...The Government also recognises that it has a role, and it exercises that role strongly. Through a range of programmes, it seeks to ensure that the market operates effectively, recognising not only the economic importance of energy efficiency but also the social and environmental benefits. Where necessary the Government intervenes in the market to achieve its objectives (11).

This brief review of the policy stance leads to two broad conclusions. First, it is most difficult even for the closest observer to identify the Government's strategic objectives in the energy efficiency area, certainly in a manner which might permit auditing or quantification of the performance of its policies in specific sectors. Rather, the Government's energy efficiency policy could more aptly be described as a clutter of expedients which (at least in the context of environmental policy), it has claimed it wishes to avoid (12).

Second, given the considerable effort which has been devoted to identifying the many barriers to the operation of market forces in energy efficiency (13), particularly in the domestic sector, it is touching that the Government - perhaps for ideological reasons - continues to place primacy of emphasis upon, and continued unshaken belief in the efficacy of, such market mechanisms.

Market mechanisms, particularly higher energy prices, cannot be overlooked as a basic stimulus for energy efficiency responses. However, many barriers, especially in the domestic sector, have been evident for a very long time and a much more proactive role is required from Government if they are to be overcome more successfully in the 1990s than over the past two decades.

(d) Specific domestic sector energy efficiency measures

Current energy efficiency initiatives towards the domestic sector in which the Government is involved, directly or indirectly, include:

- Building Regulations, and current proposals to extend them further (though not before 1993 at the earliest);
- energy labelling of buildings, although it is far from clear that the Government sees itself doing more than sponsoring the rival National Home Energy Rating (NHER) and Starpoint schemes - which the Energy Select Committee has described as two incompatible ways of telling the time;
- the Home Energy Efficiency Scheme (HEES) which was launched in 1990 and is intended to increase the uptake of basic insulation measures in low income households by means of grants and advice (14);
- the Estate Action Programme, as the main instrument for energy efficiency improvements in local authority housing;
- provision of information, although the budget for information has been much reduced;
- appliance labelling and efficiency standards, although the Government's preference is for a voluntary scheme rather than the mandatory ones beginning to emerge in draft from the European Commission (15);
- underlying research, such as that undertaken by the Building Research Establishment (16).

To these should be added the work of numerous directorates of the European Commission, including the SAVE programme and specific measures such as the draft Directive on boiler efficiency standards (17). It appears likely that, in the absence of significant new initiatives from the British Government, the European Commission is likely to emerge as the prime mover in the formulation of more vigorous energy efficiency policies.

The Regulators (Offer and Ofgas) are beginning to make some progress in propelling their respective industries further in the energy efficiency field (18). Many commentators were saddened by the Government's failure to exploit the opportunities provided by privatisation of the gas and electricity industries to ensure that much greater emphasis had to be placed by them on a range of energy efficiency issues. In addition, the powers of the regulatory bodies in the energy efficiency field, as set out in current primary legislation, leave very much to be desired.

The activities of numerous lobbies and pressure groups in the energy efficiency field continue to be impressive. However, like the energy efficiency industry itself, they remain relatively fragmented and they are often seen to lack the clout of the powerful, well-organised, vested interests on the supply side. Nevertheless, in recent years their contribution to knowledge through both research and advocacy has been increasingly authoritative despite the meagre resources at their disposal.

Some key issues for the next decade
1. The New Policy Imperative
A combination of pressing environmental constraints, a prospective return to significant energy import dependence, and increased competition in international (and especially Community) trade suggest that the UK urgently needs to address the significant economy-wide benefits of a high-profile, integrated and consistently applied energy efficiency programme.

Given the increased political salience of a range of environmental concerns, particularly CO_2 emissions and their contribution to the greenhouse effect, it appears likely that energy efficiency considerations will begin to dominate the energy policy agenda during the 1990s and beyond.

Conversely the supply side might well be relegated in significance. At least temporarily, the energy supply side at a global level appears to present fewer policy tripwires than in the past two decades although there is never room for complacency as regards the key strategic issues of supply security and diversity. In addition, as a result of privatisation as much as philosophical predilection, the Government seeks to give the impression that - apart from privatisation of coal and the 1994 nuclear strategy review - it has worked through its energy policy agenda, such as it was; indeed, so much so, that the Department of Energy was made redundant after the 1992 General Election.

Finally, the European Commission appears increasingly likely to control the main policy levers on the supply side. These factors suggest that a re-weighting of policy emphasis - away from supply to demand side issues - is possible in the coming decade.

2. Least Cost Planning
Much has been written in the recent past about the value of techniques which might be used to minimise the cost of supplying the energy services required by consumers. Least cost planning is much in vogue

in certain areas of North America, in particular. Its use has been advocated here by numerous commentators (19). Since privatisation of the gas and electricity industries, discussion of this issue has continued largely in the tradition of an earlier long-running, but essentially unresolved, debate about the relative cost effectiveness of investment in supply versus demand. As long ago as 1975, the former Select Committee on Science and Technology had called for a reappraisal of public sector investment criteria, arguing that:

> ... henceforth, the Government should consider the extent to which increase in energy demand should be met by investment in additional supply capacity or avoided in energy conservation measures (20).

Successive reports from the Energy Select Committee in the 1980s continued to explore this theme (21). However, following gas and electricity privatisation, the Government appears keen to distance itself from any role in the investment in supply/demand and least cost planning debates. When tackled by the Energy Select Committee on least cost planning and gas and electricity tariff policies, the Department of Energy responded by arguing tersely that:

> These are matters for Ofgas and Offer to consider and the EEO holds regular consultation meetings with them (22).

However, the broad issues concerning cost-effective resource allocation within the economy can be neither avoided nor delegated by national governments. The energy sector is perhaps the most capital intensive of any major economic sector. Hence here, above all, it seems important to avoid unnecessary investment in either the supply or demand sides. Evidence of continuing mis-allocation of investment resources between the supply and demand sides of the energy sector (which was carefully documented in the 1980s (23)) would raise major questions regarding the wisdom of placing so much reliance on market forces as the centre piece of the Government's energy efficiency policy. This is why, in the national interest, it is essential for the Energy Efficiency Office to set out the conclusions it has reached both from its own continuing monitoring and research and from its regular meetings with Ofgas and Offer.

As regards the much narrower point of the role of regulatory agencies in assessing their target industries' investment options in supply and demand, it would seem essential for the EEO to undertake and publish its own analysis of the pros and cons of international comparative experience in this field. Such a report would provide a valuable contribution to the rigour and quality of professional and public debate on this issue.

3. Institutional Reform

As was stated at the outset, the fact that demand for energy is a derived one poses some institutional problems for policy implementation. This is made more difficult by the extent to which Whitehall mandarins jealously guard their territory. The Energy Efficiency Office was created within the Department of Energy in November 1983, having evolved from the earlier Energy Conservation Division. The case for and against an Energy Efficiency Agency has been argued through on a number of occasions, but in view of the performance of the EEO in recent years, the case for an independent agency has grown even stronger.

For a time the activities of the EEO were backed by a strong Secretary of State, reinforced by the designation of 1986 as Energy Efficiency Year. This greatly assisted the profile of the EEO within and without Whitehall at a critical stage of its evolution. Inevitably, the policy priority given to privatisation over the subsequent years has tended to put the EEO in the shade.

This is much to be regretted, as this period has also coincided with a weakening in energy prices and hence lower attention to energy efficiency concerns. To be effective (given long lead times in policy initiatives and the lags inherent in capital stock rotation referred to above), energy efficiency policy needs to be conducted in a counter-cyclical way. When crisis is in the air, with energy prices high and supplies threatened, the market itself (aided by the mass media) will give the subject a high profile. It is in periods of relative calm that the efforts of Government are required to sustain interest on the part of energy consumers. Given this requirement, the Government's predilection in favour of agencies, and the transfer of the Department of Energy's remits to the Departments of Environment and Trade & Industry in 1992, the arguments in favour of establishing an independent agency in the energy efficiency field are overwhelming.

4. A National Energy Efficiency Audit

A priority for the next decade is the publication of regular reviews of national energy efficiency performance. As anyone who has worked in this field will know, it is at present extremely difficult to identify reliable evidence with which to measure the UK's overall performance in this field. Even when information does exist, it is to be found in a host of disparate sources. Thus it seems essential to synthesise this material in a single definitive source (akin to the regular Brown Books on North Sea oil and gas) and also to evaluate, in a critical style, both efficiency performance and the effectiveness of policy (24).

5. *Measurement, Instrumentation and Control*
It seems astonishing, given the enormous technological strides made on the supply side of the energy sector over the past century, that the meters available to domestic consumers would be recognisable by their Victorian forebears.

The advent of microelectronic controls and similar technologies offers the prospect of making energy efficiency fashionably high tech. Information is also the principal route to consumer sovereignty of the type espoused in economic texts! Despite numerous field trials across the major utility industries in recent years, progress has been disappointingly slow. There are grounds for suspicion, too, that the very process of privatisation of these industries may have impeded progress in this direction. Quarterly billing cycles are too infrequent to ensure that consumers are cost conscious. Likewise new meters will permit greater cost reflectivity in tariff setting. The fuel industries' Annual Reports invariably refer to the benefits to be derived from information technology expenditure within their own organisations. It is time that the wider consumer benefit deriving from deployment of such technology was given greater emphasis.

6. *Taxes and Subsidies*
Higher energy prices are widely advocated as the most effective way of promoting energy efficiency. This approach operates by drawing attention to energy costs and by improving the economics of energy efficiency investment. The elasticity of energy demand with respect to price has been analysed in countless academic studies which have revealed that long-term elasticities are significantly higher than short-run elasticities. This is the mirror image of the discussion above about the long lead times involved in capital stock rotation in the energy sector.

It is for these reasons that, in the context of the greenhouse effect debate, the case for carbon and/or energy taxes has been gaining support; however, rather more analysis is required before specific proposals are agreed. The low background level of real energy prices at present means that, were additional taxes to be considered politically desirable, they will need to be set at (or eventually reach) relatively high levels. Second, the relative merits of carbon versus energy taxes needs to be argued through more fully. (The case for the imposition of VAT on domestic fuels is a related debate). Third, taxes on an essential commodity such as energy in the domestic sector will clearly have regressive impacts on income distribution. Their imposition will therefore need to be matched with higher levels of revenue and capital support to those already on low incomes. Fourth, the imposition of carbon or energy taxes should be viewed in the round as part of wider fiscal reform. Revenues from energy-related taxes will offset the need for yield from other imposts and/or permit higher overall levels of public expenditure. Moves should be made cautiously.

In most of human endeavour, it seems that a combination of both stick and carrot is more conducive to sustained effort than either alone. Taxes have a negative feel about them. The Government's general policy stance essentially eschews offering consumers incentives, although experience suggests that incentives accelerate consumer responsiveness, even when set at relatively low levels.

The subsidy debate in the context of energy efficiency has often been rather arcane, overlain with references to the free rider effect - ie why pay those who will act in any event? (Rather than to a more deserving target, the company car driver!) Subsidies for energy efficiency at present compete with a myriad of other claims upon public expenditure. Were carbon or energy taxes to be introduced, there would be an opportunity to review judicious matching of taxes and subsidies as a means of accelerating, or directing, consumer responses. In such a review, the Treasury's obsessive opposition to the principle of hypothecation would need to be tackled as an early priority. At the same time, if used, subsidies would need to be targeted carefully - particularly to those on low incomes and without the capital resources to respond to any general rise in energy prices occasioned by higher taxes on energy.

Notes and References
1. This section of the text draws upon J. H. Chesshire, An Energy Efficient Future - A Strategy for the UK, (revised text of the 1986 Energy Efficiency Year Lecture), *Energy Policy*, Vol. 14. No. 5, pp. 395-412, October 1986.
2. Energy Paper No. 58, Department of Energy, HMSO, London, 1989.
3. Memorandum from the Housebuilders' Federation, *Energy Efficiency*, Third Report from the House of Commons Select Committee on Energy, Session 1990-91, HC 91-II, Vol II, p. 88.

4. It should be recalled, in this context, that the last White Paper was published in 1967 (*Fuel Policy*, Ministry of Fuel and Power, HMSO, London, Cmnd 3438) and the last Green Paper in 1978 (*Energy Policy - A Consultative Paper,* Department of Energy, HMSO, London, Cmnd 7101).

5. *Hansard*, House of Commons Debates, 19 December 1983, col.42, Written Answers.

6. *National Energy Efficiency*, National Audit Office, HC 547, London, July 1989, p.1, para. 2.

7. ibid, p.2, para. 7d.

8. ibid, p.37, table 6 and para. 4.14.

9. *This Common Inheritance - Britain's Environmental Strategy*, HMSO, London, Cm 1200, 1990.

10. Memorandum submitted by the Department of Energy, *Energy Efficiency,* Third Report from the House of Commons Select Committee on Energy, Session 1990-91, HC 91-III,Volume III, p. 4.

11. ibid., p.1.

12. In last Autumn's Environment White Paper (Cm 1200, op. cit.), the Government recognised the need to: ... *ensure that its policies fit together in every sector; that we are not undoing in one area what we are trying to do in another; and that policies are based on a harmonious set of principles rather than a clutter of expedients* (para. 1.6, pp. 8-9).

13. Such lists can be found, inter alia, in the following reports from the House of Commons Select Committee on Energy: (i) *Energy Conservation in Buildings,* Fifth Report, Session 1981-82, HC 401, Vol. I; (ii) *The Energy Efficiency Office,* Eighth Report, Session 1984-85, HC 87, HMSO, London; and (iii) *Energy Efficiency,* Third Report, Session 1990-91, HC 93 (Vol I-III). *National Energy Efficiency,* Report by the Comptroller and Auditor General, National Audit Office, HC 547, HMSO, London, July 1989, and *Energy Conservation in IEA Member States*, International Energy Agency/OECD, Paris, 1987, are also most useful.

14. Given that Brenda Boardman's paper addresses the theme of Fuel Poverty and the Social Aspects of Energy Efficiency, the HEES scheme is not analysed more fully here. Further information is also given in *Energy Efficiency,* HC 91, (op. cit.), Vol. I, pp. xxxvii-xxxix.

15. See *Government Observations on the Third Report from the Committee (Session 1990-91) on Energy Efficiency*, Fifth Special

Report from the House of Commons Select Committee on Energy, Session 1990-91, HC 566, HMSO, London, July 1991, pp. vi-vii, paras. 12-14.

16. These initiatives have been documented and reviewed by the Select Committee on Energy in *Energy Efficiency,* HC 91, Vols I-III

17. For example, the draft Directive entitled *Concerning the Approximation of National Legislation on the Efficiency Requirements for New Hot Water Boilers Fired with Liquid or Gaseous Fuels.*

18. For example, the new energy efficiency factor included in the tariff price formula of British Gas, set out in Ofgas Press Release No. 12/91, dated 29 April 1991.

19. See in particular Ian Brown, *Least Cost Planning in the Gas Industry,* Report for Ofgas, London, 1990.

20. *Energy Conservation,* First Report from the Select Committee on Science and Technology, Session 1974-75, HC 487, HMSO, London, 1975, p. vi and p. xxxviii, para. 71.

21. For details of numerous relevant references to this earlier debate see J. H. Chesshire, *An Energy Efficient Future* (op. cit. Reference 1 above) and J.H. Chesshire, *Investment Incentives: Have We Got the Balance Right?*, Chapter 5 in A. Harrison and J. Gretton, (Eds), *Energy UK 1986 - An Economic, Social and Policy Audit,* Policy Journals, Newbury, Berks., 1986.

22. See Reference 15 above, p. vi, para. 9.

23. For instance, in the various reports on energy efficiency by the Select Committee on Energy, cited elsewhere; and in J.H. Chesshire, *'Investment incentives: have we got the balance right?'*, in (ed.) A. Harrison and J. Gretton, op.cit., reference 21.

24. Another possible model, at least in terms of information provision, is the Domestic Energy Fact File published occasionally by the Building Research Establishment.

Summary of seminar on Economic Aspects of Energy Efficiency

Chaired by: Lord Ezra, President of Neighbourhood Energy Action
Speaker: Sir James McKinnon, Director General of the Office of Gas
Supply (Ofgas)

The seminar began with a talk by Sir James McKinnon on the economic benefits offered by energy efficiency investments.

He stressed the scale of the cost savings that could be obtained from the adoption of proven energy efficiency measures: up to one-fifth of the UK's total energy bill of £40 billion could be saved – roughly equivalent to the turnover of British Gas's gas supply business. The national scale of the savings did not translate impressively to the level of the individual consumer or business, however. 'For most of them, energy amounts to 5 per cent or less of their total spending and expenditure on any individual fuel is even smaller. Even if savings of one-fifth can be made in expenditure on gas, this therefore amounts to only tiny fractions of 1 per cent of total expenditure.'

The problem was compounded, he argued, by the traditional approach to energy in the UK as 'an overhead to be suffered rather than managed' – a perspective encouraged in his view by the price-setting methods of energy utilities when in the public sector. This tendency obscured the extent to which energy costs are controllable and can be significantly reduced. He noted the wide variations in energy consumption costs in offices, which indicated great scope for savings, and the large proportion of costs accounted for by energy consumption in specific sectors such as metals.

Controlling energy costs through energy efficiency produced immediate benefits in all industries. Energy bills might be small as a proportion of turnover but savings could boost profits considerably. 'For instance, a retailer with turnover of around £3 billion might have an energy bill of, say, £150 million – small as a proportion of turnover but possibly large as a proportion of profits. If the margin on turnover were around 5 per cent, profits would be of the same order as the energy bill. A 20 per cent saving in energy would therefore amount to a 20 per cent increase in profits. To achieve the same result by increased turnover would require a very considerable effort'. Similar considerations could apply to households, and the benefits of cost savings through greater energy efficiency were particularly important for low-income families for whom energy expenditure takes up a higher proportion of the household budget.

He then turned to the barriers to securing these economic benefits, and emphasised his conviction that the free operation of the market was the key to overcoming them: 'There are limits to this market... but if we forget that it is both the starting point and the main road forward, we have lost the battle bfore it has even started'. The gross inefficiencies in energy use of the former Communist command economies in Eastern Europe and the ex-USSR stemmed from the lack of any price mechanism and the consequent treatment of energy as simply an overhead, 'in practice free'. In the UK, 'we have suffered to a lesser degree from the same problems – choice for consumers has been very limited; the market has been dominated by powerful supply monopolies; people have not seen energy as a resource they can manage effectively'. In a more competitive energy market suppliers must demonstrate to customers that they can offer better value than their competitors: 'Part of this value for money will include the greater efficiency of use and this will therefore be a key element in a competitive market, supported by the proper pricing signals, and creating the incentive for energy saving investments'.

The energy market had certain limitations in relation to energy efficiency. Sir James said that he saw his role as a regulator as requiring him to act as a substitute for competition for British Gas where no market competitors existed. 'Energy efficiency is one very important aspect of this role. One of my tasks is to make British Gas treat energy efficiency as seriously as it would if it were exposed to competition'. Ofgas could play a part in tackling a number of the obstacles to greater investment in energy efficiency.

First, it was clear that the gas tariff did involve a potentially harmful distortion. Ofgas' review of the British Gas tariff formula had shown that the system gave an incentive to sell more gas rather than encourage energy efficiency. The review did not conclude that the tariff structure should be changed in order to remove all incentive to boost gas sales, given the environmental benefits of gas over other fossil fuel sources, but 'It was clear that British gas was not in a position to treat energy efficiency and energy supply on level terms'. Accordingly energy efficiency considerations were built into the revised tariff structure in the form of the 'E' Factor, which has two purposes:

- to provide British Gas with an incentive to engage in least cost planning in the sense of comparing demand-side and supply-side alternatives;

• to remove any distortions that might affect that comparison.

Sir James noted that while the E factor did not place any requirement on British Gas to engage in least-cost planning, it did provide a commercial incentive to do so. He argued that the E factor could also help tackle the problem of capital availability for energy efficiency investments by low-income householders, by making capital investment schemes targeted at such consumers possible.

A further key obstacle, he argued, was the 'huge information gap' that restricted consumers' awareness of the energy they were using and how much it was costing them. Better metering systems could improve the availability of relevant information on gas consumption and costs, and hence improve the working of the market. 'If it could be shown that British Gas was inhibited from taking action because the introduction of more effective metering would result in lower energy consumption, I would certainly wish to consider the implications for the tariff formula'. Ofgas had also sought, in its agreement with British Gas on a Code of Practice, to improve the quality of energy efficiency advice services, including energy labelling at the point of sale of appliances, and of staff training. A key issue for the gas industry was the need to raise awareness and availability of condensing gas boilers, which offer high levels of efficiency for gas central heating, and for which take-up is still low.

Discussion

The ensuing debate centred on the extent to which the competitive energy market could deliver energy efficiency improvements. Lord Ezra, chairing the meeting, recalled that he had made efforts to put energy efficiency higher up the policy agenda at the time of public discussion on the possible privatisation of the gas and electricity industries. Were regulatory bodies such as Ofgas inhibited in promoting energy efficiency by a lack of powers provided by the legislation governing them?

Sir James replied that he had no interest in expanding the powers of the regulators, and would prefer to see incentives for greater attention to energy efficiency to emerge through the workings of a freer energy market. He acknowledged that it had been difficult for Ofgas to find ways of fulfilling its duty to promote energy efficiency, and its role had been limited largely to exhortation, for example in producing information to encourage the installation of condensing boilers. The new E factor in the gas tariff formula represented the major opportunity for increased investment, but the intention was that the suppliers and manufacturers would take the lead, not the regulator. 'I do not believe you can regulate people to compete' – the key incentives would be based on perceptions of commercial interest.

Several questions focussed on the scope for promoting least-cost planning through the regulatory system, since this would not be taken up unless the utilities were given incentives to invest in it and become suppliers of 'energy services', including energy efficiency, rather than sellers of fuel. Sir James stressed that the responsibility for investment decisions lay with the utilities: it was not up to the regulators to seek any influence over such decisions, and Ofgas could not change the gas tariff structure unilaterally. The purpose of the E factor was to provide an incentive to British Gas to commit resources to investment in energy efficient generation. Ofgas would take a flexible approach to appraising the cost-effectiveness of schemes proposed for implementation by British Gas within the E factor framework.

In response to questions about the imperfections of the energy market in terms of consumer information and availability of capital for household and business investment in energy efficiency and conservation, Sir James acknwledged that there had been only 'limited acceptance' so far of energy efficient products and related services. Nonetheless, he concluded by insisting that the more highly developed the free market in energy became the more progress would be made in promoting energy efficiency.

Social aspects of energy efficiency

Dr Brenda Boardman
Environmental Change Unit,
University of Oxford

Social aspects of energy efficiency range from the effect on individual households - the social consequences - to the impact on social policy at government level. This paper covers both aspects, but is mainly concerned with the social effects of energy *inefficiency*.

Social consequences of energy inefficiency - fuel poverty

Defining poverty is a largely subjective exercise; however for the purposes of this paper, eligibility for one of income support, housing benefit, community charge benefit or family credit is taken as an indicator of poverty. All four benefits are means-tested. These 'passport' benefits give access to the main form of assistance with energy efficiency measures, the Home Energy Efficiency Scheme (HEES). The then Secretary of State for Energy stated in 1991, in relation to eligibility for HEES:

> ... there are about 7 million households in Great Britain which are ... the area in which we ought to be looking (1).

When Northern Ireland is included the number of households in receipt of the passport benefits in the UK is 7.2 million. Of these 7.2 million families, only about 100,000 have homes that they can afford to keep adequately warm (2). The remaining 7.1 million households live in homes that are poorly insulated with inadequate or expensive heating systems, so that the claimants are unable to afford sufficient warmth on their present income: they are in fuel poverty.

In 1991, there were 22.5 million households in the UK. Consequently, over 31 per cent of households are in fuel poverty. Other definitions and approaches have confirmed the 30 per cent figure (3). We are, therefore, looking at a problem that affects at least

7 million households in the United Kingdom in 1991 - over 30 per cent of all households. Fuel poverty affects a large proportion of British families.

The receipt of a means-tested benefit indicates that the claimant has limited capital. In a recent Government survey, 85 per cent of the poorest households in this country had less than £500; over two-thirds had no capital at all (4). Those that have some savings are most likely to be elderly people, who often see their money as a funeral fund.

Even if they had capital, a further constraint on 68 per cent of the poor is that they live in rented accommodation (5), with few legal rights to alter the building fabric. Another 27 per cent are predominantly older people who own their home outright, having paid off the mortgage. Thus, most of the poor have low incomes, minimal capital and landlords.

The poor are pensioners, single parents, unemployed, sick and disabled, many of whom are in the home in need of 13 hours of warmth each day in winter to be comfortable and healthy (6). Someone at work, where heating is paid for by an employer, needs to heat the home for only 7 hours a day. The poor, therefore, often need more hours of warmth than other, better-off people.

Many British homes are fairly cool in winter, in comparison with the temperatures achieved by our European neighbours. The day-time average of 16°C in UK homes masks a wide range of measured temperatures with low-income, non-centrally heated homes being 3°C cooler than those belonging to families with higher incomes and central heating (7). Temperature measurements were taken for the English House Condition Survey in the winter of 1986-7, which included a particularly cold January. The report of this energy survey was finally published on 2nd October 1991, and makes grim reading (8). When it is below -1°C outside, over 20 per cent of households have a temperature of less than 12°C inside the house (9). There is a real risk to health at this temperature.

As a result of being cold, many low-income homes are rife with condensation and the resultant mould. In 1986, 20 per cent of English homes had some condensation or damp, but for privately rented accommodation the figure was 36 per cent (10). Condensation, mould or damp existed in 28 per cent of homes in Glasgow and 23 per cent in Liverpool in 1989. Condensation means more than pools of water on the window sill. It generates foul smelling mould which may cause asthma, allergies and bronchial diseases (11). Any household forced to live in these conditions is already experiencing a polluted environment that can only be cured by adequate levels of warmth. The

Department of Health informed a recent cabinet committee that illness resulting from condensation in the home cost the health service £800 million p.a. This is one of the first examples of Government identifying the economic cost of social problems resulting from poor quality housing.

Because of their inability to afford adequate warmth, many thousands of poor people, in all age groups, will die every winter from cold-related illnesses. Every 1°C drop in temperature below the winter average will result in an extra 8,000 deaths (12). This rate of excess winter deaths is up to three times greater than occurs in countries, such as Canada and Sweden, where the winter is much more severe than in Britain, but the homes warmer (13). In two weeks in February 1991, when the weather was extremely cold, there were an additional 4,000 deaths, above the normal winter increase (14). There are no published data on the increase in illness during cold weather, nor have the substantial costs to the Health Service been quantified.

For other families, the problem of fuel poverty is evidenced by problems in paying for fuel. The desire or the need for warmth may override careful budgeting and result in a level of expenditure that cannot be afforded. Many of those with fuel debts have young children or illness in the family, or have suffered a sudden drop in income, for instance because of unemployment. Even where the manifestation of financial difficulty is as extreme as disconnection, the root cause is an inability to afford adequate warmth. The fuel poor may be cold or in debt and in many cases they are both.

Expenditure facts

In 1989, the average low-income family spent £8.36 on fuel each week - nearly 10 per cent of the household's total expenditure (Table 1). In other families, more money went on fuel (£11.53), but this represented less than 5 per cent of all expenditure. Thus, the poor spend twice as much as a proportion of income, but less in absolute terms, than better-off families. The large slice of the weekly budget that goes on fuel in low-income households demonstrates that it has a high priority for them and yet they still cannot afford to purchase sufficient. Fuel is a basic necessity and adequate warmth is a fundamental human need, but it is only available to the poor in Britain on a restricted basis. It follows that, whilst warmth may be a primary concern, the inability to purchase sufficient fuel means the poor are also deprived of other energy services, including lighting, hot water and the use of appliances.

Table 1: Household expenditure on fuel by income, UK 1989 (£/week)

	30 per cent of households with lowest incomes	70 per cent other	Average
Gas	2.83	4.63	4.09
Electricity	4.18	5.74	5.27
Coal and coke	0.84	0.67	0.72
Other	0.51	0.49	0.50
Total fuel expenditure	£8.36	£11.53	£10.58
All household expenditure	£87.28	£283.05	£224.32
per cent on fuel	9.6	4.1	4.7

Source: Based on Family Expenditure Survey, 1989, Table 5

A major problem is to establish what level of expenditure in their existing homes would provide the occupants with adequate warmth. For a typical low-income pensioner couple, living in council accommodation, an Energy Efficiency Office publication (15) gives weekly expenditure during a 30 week heating season as:

present expenditure on heating	£6.65
expenditure required for adequate warmth,	
a) existing poorly insulated home,	
with a poor heating system	£16.15
with a modern efficient heating system	£10.35
b) well insulated home,	
with a modern efficient heating system	£5.65

It is only the last combination, of good thermal insulation and an efficient heating system, that enables these pensioners to have adequate warmth for less than they are paying at present. For a low-income family, adequate warmth is only affordable in an extremely energy efficient home.

Where the house is poorly insulated (eg no cavity wall insulation), the cost of heating, for this typical couple, ranges from an extra £3.70, if there is modern central heating, up to an extra £9.50, with individual fires. In the latter case, the couple would have to more than double their expenditure to have adequate warmth. Research in Glasgow found that in small, two-bedroom flats, families would need to spend £17-22 per week on fuel to have adequate warmth and other energy services (16) - well over double their present expenditure on fuel.

The above example is for a pensioner couple on a low-income in local authority accommodation: this group represents about 25 per cent of all the fuel poor. There are insufficient comparable data for other groups of the population. However, local authority dwellings are marginally better insulated and easier to heat than privately rented accommodation, partly because they have been built more recently and to a higher standard. Approximately half of all low-income households (across all tenures) have some central heating, but rarely a modern, efficient system. Therefore, the majority of the fuel poor come into the category of needing to spend an extra £9.50 per week on heating. A minority would need spend as little as £3.70. For these reasons, an average of £7.50 has been taken as a conservative estimate of the extra expenditure needed by low-income families if they are to keep warm in their present homes. Based on a 30 week heating season, this gives:

£7.50 x 30 weeks x 7m households = £1,575 million each year

This is the additional money required to provide adequate heating only. Extra benefit payments would also be needed to ensure that the household had a satisfactory supply of all the other energy services. But this has proved impossible to quantify and has been ignored here. It is worth noting, however, that the £7.50 per 30 weeks is equal to £4.30 per week throughout the year, which would bring the fuel expenditure of low-income households above that of other households (Table 1). This is not surprising, because the fuel poor do need more energy to obtain the same standard of energy service in their energy inefficient homes. In addition, they may have a higher demand for energy services, because of their greater occupancy of the home.

To provide adequate warmth through additional income, claimants could be given the extra money either through cash benefits or fuel vouchers. Using fuel vouchers would ensure that the household spent the money on fuel. This additional expenditure would be equivalent to an increase in national domestic fuel expenditure, consumption and pollution of about 15 per cent.

If the money is given as cash, because of the considerable demands on household expenditure in claimant families, only a proportion would go on fuel, perhaps 10 per cent. Thus, to raise the level of energy services in low-income homes through additional income support would cost several billions of pounds, every year, indefinitely.

Worse still, all of the money would be going to subsidize the inefficient use of energy in poorly insulated homes with expensive heating systems. Carbon dioxide emissions would increase substantially, particularly because electric space and water heating are concentrated in low-income homes. To provide low-income families with sufficient money to keep warm in their existing homes would be phenomenally expensive. Additional income as a solution to fuel poverty is neither economic nor environmental good sense.

Income details

Income support scale rates are not based on an analysis of the cost of achieving a minimum standard of living and none has been specified by government since Beveridge in 1948 (17). The theoretical amount for fuel per client group cannot be identified. The DSS does not know how much claimants have for heating (or energy) out of benefits, it does not know that this is an adequate amount, but it assumes that it is.

Since April 1988, the Department of Social Security has adopted a policy to provide income support based on social characteristics, but not on variations in energy efficiency. All members of the same client group are assumed to have homes of equal energy efficiency and to pay the same for fuel. The DSS does not recognize that the cost of warmth can vary between households and certainly not that it may vary by a factor of ten (18).

Housing costs are excluded from basic benefits, because it is recognised that the cost of housing varies substantially between one claimant and another, with a factor of two cited as the range of costs between comparable dwellings. Therefore, 'to include housing in the scale rate would leave many householders with an inadequate amount for other needs' (19). The parallel logic of variations in the cost of energy services is not recognized by the DSS or the Treasury.

Government policies have exacerbated fuel poverty, particularly through the actions of the Department of Social Security. The Social Security Review in 1988 abolished Heating Additions, though restrictions on eligibility had started in August 1985. Heating Additions worth £500m a year were being paid to 2.7m claimants in 1988. The bulk of these payments, about £400m, was paid on the basis of social characteristics, usually age. In theory, this money was incorporated into the new Income Support scheme.

The remaining £100m was based on the physical characteristics of the dwelling, the heating system or the fuel used and claimants were not compensated for this loss. The Social Security Review also required claimants to be responsible for some of their community charge and all water rates. In an analysis of the cumulative effects of the various changes, the National Right to Fuel Campaign established that those who had lost the most were households with energy inefficient homes (20). Those who were known to have homes that were expensive to heat were penalized most heavily under the Social Security Review. Under the current system, claimants have no entitlement to regular payments for heating costs nor for financial help with problem fuel bills.

Social fund

The Social Fund is intended to provide assistance in exceptional circumstances, over and above regular benefit entitlement. One of three mandatory grants under the Social Fund is for Cold Weather Payments - a measures that comes sharply into focus during periods of exceptionally cold weather, such as that of February 1991. Only about a third of the fuel poor are eligible to claim (2.2 million last winter) and of these about a third actually claimed and received their £6 per week. The scheme cost a total of £8 million in the 1990-1 winter, excluding administrative costs, which are known to be high. The payment was, therefore, about £6 for 2 weeks for 0.7 million households in comparison with the actual need: £7.50 for 30 weeks for 7 million households.

A revised scheme was announced in June 1991. It features several improvements, particularly that it will be paid automatically to an eligible claimant and can be based on predictions of severe weather. This resolves two of the main criticisms - that it had to be claimed retrospectively. The maximum payment stays at £6, which, as shown above, is inadequate according to the EEO's own calculations during average winter weather and will certainly be so when it is below freezing. The total number of eligible households is still only 2.6 million, out of the 7 million claimants in receipt of passported benefits and in fuel poverty. Cold Weather Payments are designed to provide emergency help and that is all they can do. They will continue to be desperately needed, because of the inadequacy of the basic rates for those in energy inefficient housing: cold weather affects people in poor housing disproportionately.

Prior to April 1988, claimants could obtain single payments to cover items of exceptional expenditure, such as household equipment.

These grants have now been replaced by, in most cases, loans from the Social Fund. There are considerable constraints for applicants to the Social Fund and these are resulting in claimants being refused loans, or only being offered a part loan even for essential equipment, such as heaters. In many cases, a claimant will refuse a loan, because s/he cannot afford to pay it back (21).

Many low-income households have heating and electrical equipment that is in poor condition or not working. In a survey of 25 homes, 8 per cent of appliances were faulty or broken and 42 per cent over ten years old (22). Even when a loan is available, it often permits the purchase of second-hand equipment only. Thus, the operation of the Social Fund is increasing fuel poverty in many aspects of fuel consumption, because it prohibits the purchase of modern, efficient appliances (23). This is particularly true of appliances using electricity.

There are no programmes at the moment that assist the fuel poor to use energy efficiently in 'white' or 'brown' goods - refrigerators, cookers, washing machines, TVs - and the only policy to assist with water heating is the provision of tank jackets. Two-thirds of domestic electricity is used for cooking and appliances (24) so it is extremely important that electricity is used efficiently in these appliances. The HEES Network Installers or the utilities could be funded by the Social Fund to undertake regular safety and efficiency checks on the appliances in the homes of claimants and to provide good quality, efficient replacements where appliances are dangerous, faulty or expensive to run.

Government expenditure on energy efficiency for the poor

The root cause of fuel poverty is the energy inefficiency of the house. The inefficiency results from a lack of investment in thermal insulation measures to improve the building fabric, as well as a dependence on expensive- to-run heating systems and appliances. The price of fuel contributes to the high cost of energy services obtained by the poor and their plight is exacerbated by low income, but the main problem is the lack of capital investment in energy efficiency. The poor have to buy high cost warmth - that is what living in an energy inefficient home means.

About £300 million of government money has been invested in low-income homes over the past 12 years (25) - an average of £25 million p.a. This has been spent almost exclusively on loft insulation and draught- proofing. Yet, only 33 per cent of households in socio-economic groups D and E have any draught-proofing and 83 per cent have some loft insulation (26). These figures refer to proportions of dwellings where the measure could be incorporated and do not imply an adequate level of provision. For instance, the loft insulation could be as little as 25mm, whereas 150mm is required to comply with the 1990 Building Regulations. After 12 years of grant aid, the majority of the poor still do not have even the most basic of insulation measures.

The Government admits that 'low-income households ... occupy many of the least efficient homes' and is giving priority to these through the EEO's Home Energy Efficiency Scheme (27). Despite this focus, the Government still require the fuel poor to contribute to the cost of the work (up to £15) and this is a disincentive for those most in need.

The HEES provides basic insulation measures (loft insulation, draughtproofing, pipe and tank lagging) and £10 worth of advice to low-income homes. The budget is £26 million for 1991-2, and a projected £32 million for 1992-3. This level of grant is below the peak of £60 million claimed by Cecil Parkinson when Secretary of State for Energy (28). At the same time, the aim is to double the rate of work undertaken to 200,000 jobs per year, despite acknowledgement by the Secretary of State for Energy that there are 1.7 million households without adequate loft insulation and 4.6 million houses with no draught-stripping, who are thought to be eligible for assistance under HEES (29).

The legislative basis for HEES is the Social Security Act 1990, which permits a wide range of energy efficiency measures to be undertaken for the benefit of low-income homes. It is one of the ironies of fuel poverty policy that the greatest hardship is being caused by the income support policies of the Department of Social Security, but it is also DSS legislation that is providing the basis for capital investment by other departments.

The Local Government and Housing Act 1989 contains provision to pay Minor Works grants to certain householders in receipt of one of the passport benefits. This legislation emphasises assistance with thermal insulation, but is restricted to owner-occupiers and tenants renting privately in England and Wales (in total a maximum of 35 per cent of the UK poor). Only the elderly can obtain assistance with inadequate heating systems - often the most cost-effective energy efficiency improvement. Local authorities have discretion as to whether they provide any money at all for Minor Works Assistance. During the system's first six months, 4,800 grant applications were received for an average of £620 (30). If all the grants were approved, this would represent some £3 million worth of work. There is currently no information about the numbers of local authorities that are offering this grant and, therefore, its national coverage.

The final Government initiative of relevance is the Green House programme in England, which concentrates on improving the energy efficiency of blocks of local authority dwellings that are structurally sound. The budget is for two years only: £10 million in 1991-2 and £50 million in 1992-3. Thus an important feature in the Government's proposals to reduce carbon dioxide emissions applies to England only and ends in April 1993. Since just over half of all local authority housing is occupied by low-income households, only about 50 per cent of the Green House programme expenditure is directly assisting the fuel poor.

Total current and projected Government expenditure on improvements to the energy efficiency of low-income homes is:

£35 million in 1991-2, through HEES, Minor Works and Green House Programme;

£65 million in 1992-3, through the same programmes;

£50 million in 1993-4, through HEES and Minor Works only.

Utilities and regulators

Investment in the more efficient use of energy could have been required of the gas and electricity industries when they were privatised. Instead, despite vigorous attempts by MPs and Lords, the Government only required the industries 'to promote' energy efficiency. Nor do the pricing formulae permit the pass through of any investment costs. Both these industries, therefore, were established with a profit basis predicated upon growth in energy sales. Not only does this mean that the industries have not chosen to invest in energy efficiency, it also means that they are strong advocates of increasing energy consumption.

For British Gas, this situation is expected to change from April 1992. The Director General of the Office of Gas Supply, in reviewing the gas tariff, is proposing that British Gas should invest directly in the more efficient use of gas, particularly in low-income households, and that this cost can be included as an allowable expense: the E factor. Hopefully, Ofgas will make a certain level of investment a requirement and will be fairly stringent in its definition of eligible work. Then, substantial new schemes to aid the fuel poor with capital investment could result.

In addition, and partially overlapping with the E factor requirements, the Director General of Ofgas, Sir James McKinnon, has proposed that British Gas should set up a trust fund worth £50 million 'to attack the problem of fuel poverty'. Part of the money would be raised from consumers and matched by British Gas (31).

Similar advances by the Office of Electricity Regulation offer are not seen as imminent. Each of the public electricity suppliers has been required to produce a Code of Practice on Energy Efficiency. Offer is proposing to monitor how effective these Codes are at promoting energy efficiency before considering whether to use the available powers to set standards of service or to include an E factor in the electricity tariff. The Codes of Practice concern the provision of verbal and written advice and do not involve any direct capital investment in the more efficient use of electricity.

Energy prices and carbon dioxide emissions

The cost of domestic fuel, at the meter, can vary by as much as a factor of four, with general tariff electricity costing four times as much as gas. Even when the efficiency of the appliance is taken into account, there is still a ratio of 3:1 in the cost of useful energy from an electric bar fire in comparison with gas central heating. Any household using on-peak electricity for heating is, therefore, buying the most expensive warmth. At least 1.8 million low-income households depend on this form of heating or have problematic electric central heating, such as underfloor or ceiling heating (32). This does not include people with the cheaper, night storage heaters. No Government policy has ever grant-aided fuel switching in individual households with expensive electric heating, although this would be the most cost-effective way to reduce fuel poverty.

The amount of carbon dioxide emitted per unit of fuel used also varies by a factor of four in exactly the same way as price; electricity is four times more polluting than gas. This is true for all types of domestic electricity, whatever time of day it is used. Thus policies to assist fuel poverty through fuel substitution are also the most effective at reducing carbon dioxide emissions in the domestic sector. The efficient use of electricity in domestic appliances, referred to in the section on the Social Fund, is also vital.

Most of the proposed solutions to global warming involve raising fuel prices in order to cut consumption - an approach that would increase further the incidence and effects of fuel poverty. Conversely, fuel poverty would be reduced if the poor could afford to use more fuel and be warmer, but this would increase the emission of greenhouse gases. The clear conflict between these two approaches can be avoided if policies are based on the more efficient use of energy, through direct investment in low-income homes.

Considerable research has been undertaken into the effects of carbon taxes or similar fuel taxes, including value added tax. Work by the Institute of Fiscal Studies has shown that a 15 per cent increase in price would result in a 10 per cent drop in consumption by the poorest households, but only a 2 per cent decline in the richest quintile (33). This confirms that real increases in the price of fuel are particularly regressive for low-income households.

Proposals for a combined carbon and energy tax are being discussed within the European Community and incorporate fiscal neutrality - the tax raised from energy consumption is offset by tax cuts elsewhere (34). If the rebate is made through Corporation Tax, it benefits industry and commerce only. The domestic consumer in this situation would end up subsidizing business. If the rebate were through income tax or value added tax, the low-income consumer would be contributing to a fund for the better-off. Such an energy tax would be the worst possible scenario for the fuel poor.

Even if the additional revenue raised by energy taxes were used to bolster benefits for the poorest households, this would only be a compensatory policy attempting to maintain the status quo. The additional benefits would have to be of considerably greater value than the extra cost of the energy taxes, in order to maintain energy expenditure levels in low-income households. This is for two reasons. First, the higher priced fuel has a lower utility for the consumer: it 'is not worth buying' so much. Secondly, householders only spend a proportion of their income on fuel and would use some of the additional money for other purchases, unless the benefit was given as fuel vouchers. Therefore 'redistribution of the tax revenues may be inadequate, and a package of measures, perhaps including reduced energy costs for the vulnerable elderly or measures to improve the heat efficiency of their homes, may be required' (35).

Direct investment is required, and not just in the homes of elderly people. Only about half of the fuel poor are elderly, so that policies to minimize hardship should be based on the passport benefits, to assist all 7 million households in all age groups. As the Energy Select Committee has recommended:

Any additional taxes on fuel and power should be conditional both on a much larger programme of energy efficiency investment for low-income households and on increased social security payments to protect the low-income households occupying the least energy efficient housing (36).

A programme for affordable warmth

To bring all low-income homes up to a standard which provides cleaner, affordable warmth will take a long time. The rate of construction, particularly in the public sector, has declined, so most new housing is in the private sector and not accessible to the poor. As unwanted houses are being demolished at a rate of 12,000 p.a, it will take 2,000 years to replace the present housing stock. The majority of families will, therefore, only benefit from the renovation and upgrading of existing properties. It has been proposed that a reasonable target for an energy efficiency programme in low-income homes would be:

500,000 low-income homes upgraded each year .... *This will ensure that by 2005, all of the poor are living in homes that they can afford to keep warm and which create less pollution* (37).

The proposed standard, equivalent to the 1990 Building Regulations, together with double glazing and gas central heating, is rarely achieved in improvements for low-income families. This is why it is possible to state that the vast majority of the poor are in fuel poverty - the lack of energy efficiency in British homes ensures that this is the case.

The cost of this programme would average £2,500 per house, assuming certain economies of scale. The total cost would be £1,250 million each year and £17.25 billion over the whole period. The Government's investment in HEES, an estimated £32 million in 1992-3, is, therefore, less than 3 per cent of that needed for an effective attack on fuel poverty.

It is worth emphasising the size of the gap between the Government's present Home Energy Efficiency Scheme and a programme for affordable warmth. The HEES is only covering the most basic insulation measures: draught-proofing, loft insulation, tank and pipe lagging, together with a modicum of energy advice. The stated doubling of activity will still only mean that 200,000 homes per year are treated with these measures. At this rate of work it will take over 20 years to provide only the most basic improvements to all low-income homes. Draught-proofing materials are not guaranteed to last more than 10 years - a Forth Bridge scenario of never-ending work just to draught-proof the homes of the poor.

The inadequacies of the present scheme could be forgiven if there was evidence that the Government proposed to build upon the existing scheme as quickly as possible. However, no new initiatives have been proposed to date. Potential additional measures could include cavity wall insulation and heating system installation, requiring a considerable preparation time for Network Installers to achieve the appropriate level of training and to invest in new equipment. There is no evidence that the Government is proposing anything beyond the limited HEES that exists at present.

As is clear, to complete the proposed programme in 14 years will require strong political determination to deal with the problem of fuel poverty. Even with this commitment, most poor families will have to wait several years before their home is improved. Meanwhile if they are not to go cold, families in untreated homes must be given more money in order to keep adequately warm. Although this will be both expensive and environmentally damaging, it is necessary, given the extent of fuel poverty and the legacy of hard-to-heat housing in Britain. The additional yearly expenditure on benefits and the resulting extra pollution should provide the incentive to keep the capital expenditure programme on target.

As we noted earlier, the total additional income support required to provide adequate heating for the fuel poor in their present homes has been calculated at £1.575 billion per annum, based on conservative estimates. This level of support diminishes as the capital programme is effective and low-income homes are improved. Government expenditure, therefore, will be a combination of this revenue expenditure and the £1.25 billion annual capital expenditure, which does not reduce until all homes are treated (Figure 1). When the investment programme is complete, say in 2005, both forms of expenditure should cease.

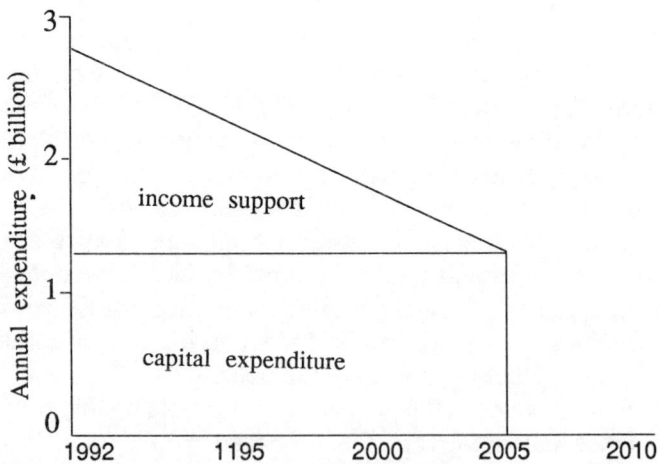

Figure 1 Government expenditure profile to provide affordable warmth in UK

The National Right to Fuel Campaign has proposed that a potential source of funds for this programme is the Fossil Fuel Levy on electricity sales (38). The Levy is paid by all consumers and represents about 10 per cent of the average household electricity bill, or £30 a year. At the moment, the money is used to subsidize the cost of nuclear power and renewable energy, in order to meet the social objective of increased security of supply. The European Community requires the Levy to be halved by 1998 and then for it to cease. The amount raised is £1.15 billion a year. The proposal is that the Levy continues at the same level, and is not decreased, with the additional money used to fund the alternative social objective of more energy efficient low-income homes. This would require the European Commission's approval.

An added benefit of using the Fossil Fuel Levy would be that all electricity users contribute to a fund for the benefit of the fuel poor. This is effectively the same basis as the E factor in the Ofgas gas tariff review, so that both utilities would be contributing to the more efficient use of energy at a significant level. This direct investment approach would have the opposite effect to the European Community plan for carbon taxes (see above), as it would benefit the fuel poor, not penalize them further.

Benefit levels could be related to the energy efficiency of the dwelling through energy audits. These vary in sophistication and expense, but are increasingly computer-based. There are two systems of domestic energy audits in the UK at the moment, the National Home Energy Rating (NHER) and Starpoint. Once an audit has been completed it can provide:
- the basis for energy allowances to claimants and, thus, protection against real fuel price increases;
- the method of prioritising the worst housing;
- the information for prioritising work within a house;
- the basis for rental levels.

If rental levels were linked to the energy efficiency of the dwelling, through an audit, this could incorporate landlords as a source of funding. The investment would be stimulated by making rents reflect the cost of keeping warm for the tenant, so that the occupant would be compensated if a property were expensive to heat and the landlord would have a financial incentive to improve the dwelling. After energy efficiency improvements the property would attract a higher rent, thus paying off the landlord's investment.

If landlords and the utilities are involved in the energy efficiency programme, in the way outlined, very little of the required investment needs to come from central Government. Even this residual expenditure would have the advantage of creating employment and thus reducing Government expenditure on unemployment benefits.

The Energy Select Committee has suggested a further link between the fuel industries and the Department of Social Security:

In consultation with OFGAS and OFFER, the Government should negotiate with the utilities a programme to assist energy efficiency investment among low-income households in return for the benefit the utilities derive from the Department of Social Security's *fuel direct* system (39).

'Fuel direct' involves the repayment of a fuel debt together with current consumption administered by the DSS through a direct deduction from the claimant's benefit at source. The DSS is effectively acting as a debt collection bureau to assist claimants who have difficulty in budgeting. This service is of considerable assistance to the fuel utilities, who should respond to this benefit by providing resources to make some of the poorest households more energy efficient.

Neighbourhood Energy Action and other caring agencies, supported by the Department of Social Security, proposed distributing leaflets advertising the Home Energy Efficiency Scheme to any claimant enquiring about fuel direct. This proposal foundered because the Energy Action Grants Agency (EAGA) decided there was a potential lack of funds. The DSS advises 300,000 households each year about the possibility of going onto fuel direct, whereas EAGA has sufficient grants for only 200,000 households and did not wish to promote a service it could not guarantee.

It would be eminently sensible to target energy efficiency measures on those known to have such difficulty in obtaining affordable warmth that they get into debt with their fuel bills. This initiative is being prevented by lack of resources.

Departmental responsibility

Two departments - the Department of the Environment and the now defunct Department of Energy - have accepted that there are variations in the efficiency with which energy is used and that many of the least energy efficient dwellings are occupied by low-income households. The Department of Social Security refuses to recognize variations in the cost of keeping warm that are caused by energy inefficiency and pays benefits at the same rate for all members of a client group. This contradiction in Government policy should be addressed.

Energy efficiency policies originating from both the Department of the Environment and the Department of Energy have emphasised thermal insulation improvements. Neither has tackled the problems of fuel substitution and of grant-aiding the provision of more efficient heating systems for the fuel poor. One of the departments should take responsibility for ensuring that low-income households have access to the most economically efficient heating systems, now that electricity and gas have been privatised and are meant to be competitive.

Beyond the lagging of hot water pipes and tanks, there has been no programme to assist in the efficient use of energy in the home for purposes other than space heating. From 1979, the Department of Energy was involved in discussions on labels to indicate the energy consumption of new appliances, but minimal progress has been made. Even such a development would be of minimal benefit to low-income households who are rarely able to purchase new appliances.

In the short term - about the next 15 years - the Department of Social Security must provide much greater assistance to enable claimants to keep warm, if fuel poverty is to be reduced. This will cost over £1,500 million a year. The DSS will continue to have the largest responsibility whilst the other two departments remain so parsimonious.

Environmental and social policies have to be integrated at Government level, just as they are at the household level. The present separate and contradictory policies indicate a lack of concern for the fuel poor.

Conclusions
Some limited progress has been made in the last few years. In 1985, Peter Walker, when Secretary of State for Energy, stated that there was no such thing as fuel poverty (40). Ministers are now prepared to acknowledge the existence of fuel poverty and to devise policies to help low-income households.

Government capital expenditure on energy efficiency improvements is, however, only about 3 per cent of that needed. The severity and extent of fuel poverty has to be recognized in a programme of investment that reflects the magnitude of the problem and the hardship suffered.

The Department of Social Security administers a benefit system that is unresponsive to the cost of keeping warm. As a result there are 7 million households in the UK unable to afford adequate warmth and who are, therefore, in fuel poverty. Because they live in energy inefficient homes, these families obtain poor value for the money they spend on energy and are likely to have both high fuel bills and cold homes.

The economic and social costs of this individual hardship and suffering are rarely assessed. It is known that there are 26,000-52,000 extra deaths in winter as a result of cold-related ill health. These are premature, avoidable deaths. Recently, the Department of Health has established that condensation-related illness, created by cold, damp homes, costs the national health service £800 million a year. A true assessment of the total costs, to society and to the individual, of fuel poverty would assist in the development of accurate cost-benefit analyses and least cost planning.

The problems for the fuel poor could be compounded in future by policies to reduce environmental pollution. Fuel price increases - the market approach to reducing greenhouse gas emissions - would result in lower levels of energy consumption, but at the expense of increased deprivation in low-income homes. Additional income support for the poor would be extremely expensive and would increase pollution levels at a time when the UK is meant to be reducing them. The two policy objectives of a cleaner environment and warmer homes can only be achieved through direct investment in the more efficient use of energy.

The poor cannot afford that investment themselves, therefore the money has to come from external sources - Government, the fuel industries or landlords. A programme to provide the 7 million low-income homes with clean, affordable warmth could cost £17.25 billion, because the British housing stock is so woefully insulated and inefficient. The level of improvements needed and the number of houses to be treated, indicate that a programme to provide affordable warmth would take a minimum of 15 years to implement. By 2005, our poorest households could have a decent home environment to live in without adding to external pollution.

This is a long-term proposal. During the short-term, perhaps as little as 15 years, there will have to be additional payments through the social security system to any claimant whose house has not been improved. This will increase the carbon dioxide emissions from these homes temporarily, but provide the necessary incentive to maintain the investment programme. If this level of activity can be achieved and there are no real fuel price increases creating additional hurdles, by 2005 the problem of fuel poverty could have disappeared accompanied by a lowering of carbon dioxide emissions from this group of households of at least 6 per cent. The poor would be warm, with lower emission rates and homes that provide them with a healthy internal environment. Capital investment reduces both fuel poverty and the greenhouse gas emissions simultaneously. Nothing else has these dual benefits.

References
1. HC 91-III, *Energy efficiency,* Third Report, Energy Committee, Session 1990-91, Minutes of Evidence, HMSO, 1991, q.625
2. B. Boardman, *Fuel poverty: from warm homes to affordable warmth,* Belhaven, London, 1991, p.206
3. B. Musannif, 'Providing affordable warmth', *Energy Management,* September/October, 1990, p.9; Boardman, 1991, op.cit. pp 44-45
4. Social Security Statistics, HMSO, 1989, table 34.55
5. Family Expenditure Survey, Department of Employment, HMSO, 1989, p.6.
6. B. Boardman, *'Activity levels within the home',* paper presented to the Joint Meeting CIB W17/77, Controlling Internal Environment, Budapest, 18-20 September, 1985.
7. D.R.G. Hunt and M. Gidman, *'A national field survey of house temperatures',* Building and the Environment, vol 17, no 2, pp.107-24, 1982.
8. DOE, *English House Condition Survey:* 1986, Supplementary (Energy) Report, Depart. of the Environment, HMSO, 1991.
9. Ibid, para.7.46
10. NEA, *Fuel poverty briefing,* Neighbourhood Energy Action, Newcastle, undated, p.2
11. S. Hunt, *Are houses a health hazard?,* Energy Action Bulletin, Neighbourhood Energy Action, Newcastle. pp.8-9, January 1990
12. M. Curwen and T. Devis, 'Winter mortality, temperature and influenza: has the relationship changed in recent years?', *Population Trends* 54, OPCS pp.17-20, 1988.
13. B. Boardman, *'Seasonal mortality and cold homes',* paper given at Unhealthy housing conference, Institution of Environmental Health Officers and Legal Research Institute, University of Warwick, 14-16 December, 1986.
14. WACH, press release. *Winter of Action on Cold Homes,* London, 1991.
15. EEO, *Energy consumption guide - A councillor's guide to affordable warmth for tenants,* Best Practice Programme 2, Department of Energy, 1990.
16. B. Sheldrick and D. Joyce, *Heating costs in Easterhouse,* Heatwise Glasgow, Glasgow, 1989.

17. K. Andrews and J. Jacobs, *Punishing the poor - poverty under Thatcher,* Macmillan, London, 1990, p.178
18. Boardman, 1991, op.cit., p.223.
19. Treasury, 1985: personal communication.
20. J. Crowe, *Colder by decree,* National Right to Fuel Campaign, Birmingham. 2nd edition, 1991.
21. NACAB, *Hard times for Social Fund applicants,* E/1/90, January, National Association of Citizens Advice Bureaux, London, 1990. Child Poverty Action Group, Evidence to the Social Services Committee Social Security: changes implemented in April 1988. Ninth report, session 1988-89. Cmd 437-III, HMSO, 1988.
22. R. Sadler, *The efficient use of electricity in low-income households.* Report on qualitative research in Ipswich and Bristol, Bristol Energy Centre, Bristol, 1991, pp6-7.
23. B. Boardman, and T. Houghton (1992), *Efficient use of electricity in low-income households,* Bristol Energy Centre, Bristol
24. Boardman, 1991, op.cit., p.153
25. Ibid, p.70
26. Hansard, WA 10.5.90
27. HC 566, *Government observations on the Third Report from the Committee (Session 1990-91) on energy efficiency,* Fifth Special Report, Energy Committee, Session 1990-91, HMSO, 1991, p.x.
28. Fuel News, quarterly newsletter of the National Right to Fuel Campaign, Birmingham, Autumn 1990, p.1.
29. HC 91-III, op cit. ref.1, p.156
30. Care and Repair, July 1991, p.3
31. Ofgas Press Release, 33/91
32. Boardman, 1991, op.cit.,p.99
33. S. McKay, M. Pearson and S. Smith, 'Fiscal instruments in environmental policy' *Fiscal Studies,* Vol 11, No 4, November, pp.1-20, 1990
34. The Guardian, 26.9.91
35. McKay et al, 1990, op.cit., p.17
36. HC 91-I, *Energy efficiency,* Third Report, Energy Committee, Session 1990-91, Report, HMSO, 1991, p.1ii
37. B. Boardman, *Fuel poverty and the greenhouse effect,* NEA, FOE, National Right to Fuel Campaign, Heatwise Glasgow, 1990
38. HC 91-II, *Energy efficiency,* Third Report, Energy Committee, Session 1990-91, Memoranda of Evidence, HMSO, 1991, pp 96-8
39. HC 91-I, op. cit. ref. 36, p. 1ii.
40. Energy Action Bulletin, Neighbourhood Energy Action, Newcastle, October 1985

Summary of seminar on Social Aspects of Energy Efficiency

Chaired by: Peter Barclay, Chairman, Social Security Advisory Committee

Speaker: Rt. Hon. Tony Newton MP, former Secretary of State for Social Security; now Leader of the House of Commons.

The seminar speaker, Tony Newton MP, then Secretary of State for Social Security, observed in his introductory remarks that energy efficiency investments make sense for most people – reductions will be seen very quickly in fuel bills, and investments will be recovered generally in a fairly short time. On the other hand, he noted that people on low incomes tend not to have the capital to make these investments, and may live in homes that are very energy inefficient.

With that in mind, the Government had taken a number of steps to improve the energy efficiency of the housing stock. The EEO's budget for promoting energy efficiency in low income households had been increased in the 1991 Autumn Statement by 50 per cent to £40 million. The existing budget of £27 million already allowed for some 200,000 insulation jobs in the present financial year. This increase took no account of the monies spent by local authorities on the housing stock generally: around £300 million was spent in 1989/90 on improving insulation and other energy conservation measures; expenditure continues on this scale.

In addition there is the Greenhouse Energy Efficiency Demonstration Programme, introduced to encourage local authorities to improve the energy efficiency of their housing stock: an extra £10 million had been made available for this in the current year, and a further £50 million would be available in 1992/93. These were quite sizeable increases in the area of policy under discussion.

It was clearly a huge task to upgrade the quality of the nation's overall housing stock. However, the latest English House Conditions Survey, covering 1981-86, showed that there has been a considerable improvement in energy efficiency and insulation levels, particularly in the least energy efficient properties and among low income households.

He observed that the Secretary of State for Social Security is always faced with suggestions for additional benefits for heating, such as Brenda Boardman's proposals for benefits to reduce fuel poverty and promote energy efficiency. In considering such proposals we needed to remind ourselves of the situation before the 1988 reform of the benefit system. One of the problems with the pre-1988 system was that over the years there was a proliferation of special additions for particular items, including heating, which made the system complex and hard to understand for recipients and for administrators. One key aim of the reforms was to simplify the structure of social security in order to promote more efficient delivery of benefits.

This general aim affected policy towards the old Heating Additions. Tony Newton argued that these had two main problems, which indicated the difficulties involved in pursuing the policies sometimes urged on Government. First, many of the additions were defined with reference to client groups, such as pensioners and disabled people. Thus in practice these were equivalent to premium payments for those groups, and the reform recognised this explicitly by bringing them together in a more coherent and comprehensible system of premiums for the client groups in question. The label 'Heating Addition' had disappeared, but the money covered was included in the new system of premiums. In any case the payment of the old addition was not tied in any way to actual expenditure on fuel by the recipient.

Secondly, some Heating Additions were defined by reference to special circumstances. These proved even harder to administer, and also caused various anomalies. An addition targeted on estates was hard to operate because of the problems involved in defining exactly what was and was not an 'estate', and the central heating addition brought in early in the 1960s was outdated by technical change leading to central heating systems becoming more cost-effective. These problems at the practical level indicated some of the difficulties of implementation that need to be considered in relation to the policies sometimes proposed on fuel poverty.

Tony Newton argued that it was hard to see how these types of difficulties could be overcome any more readily than they could before the 1988 reforms. Even leaving aside the financial details of Brenda Boardman's proposals, the problems of implementation would be very great, given the need to make an assessment of some 4 million claimants' homes in some detail to gauge how energy efficient they were, and to reassess every time there was a new insulation scheme. He remained convinced of the need to focus on groups that were likely to have extra needs, which may often involve heating, by building on the system of premiums introduced by the Government and the disability benefits system to be established in April 1992. Measures had been taken to assist groups with specific needs such as help with heating: all pensioner premiums had been increased in real terms within the previous 3 years, amounting to extra benefit expenditure of over £300 million per year; and there would be a benefit increase in 1992 of 7 per cent for those on income support. So there was extra money in real terms, albeit not labelled as a heating addition, in the form of increases to the premiums made available to vulnerable groups such as the older and less well-off pensioners, the disabled, and families on low incomes.

Finally, he pointed out that the Cold Weather Payments Scheme was specifically designed to help with extra heating costs in very cold weather and was targeted at the most vulnerable groups receiving income support – pensioners, sick and disabled people, and families with young children. A package of changes to this scheme had been introduced, increasing benefits, dropping some restrictions and making payments automatic. As a result of these changes, and in particular because of the automatic triggering of payments, some 400,000 more people would be eligible. All of this was in addition to the extra help the households in question will receive from the other measures mentioned earlier.

Discussion following the Secretary of State's talk
Several of the contributions from the floor emphasised the extent to which energy efficiency is an area where policy coordination across government departments is needed if low-income households are to benefit fully from its advantages. For example, the initial comments drew attention to the widespread adoption elsewhere in Europe of district heating systems which helped countries with colder climates and dearer fuel tariffs to avoid the problems of fuel poverty. By contrast, power stations in the UK throw away enough hot water to heat every building in the country. Whilst this was not an area of direct responsibility for the Department of Social Security, questions about the prospects of a European Community Carbon tax, and in particular the likelihood of a fiscally neutral policy in this area, were more immediately relevant. One speaker feared that such an approach would be regressive in its impact, and asked what would be the DSS view on the implications of the European debate on a carbon tax.

Tony Newton said that the DSS would clearly wish to consider the policy implications of any energy tax measures. Since, however, energy costs were already included in the index used for uprating benefits, price increases would automatically flow through into benefit rates.

Brenda Boardman, author of the background paper for the seminar, said that it was not clear how much of the old heating addition money had re-emerged in the premiums system and stated that the premium system did not take energy efficiency considerations into account at all. She went on to say that the key problem with fuel poverty policy was the lack of information on the range of costs for low income households on benefit in keeping warm. Her paper's estimate of a 1:10 ratio was a guess – the data is not collected. This meant that a uniform benefit was being paid to households with non-uniform expenditure in significantly varying circumstances. The solution lay in a multi-purpose system of energy auditing, which would allow better targeting of resources, depending on real need and household conditions, and provide landlords with an incentive to improve properties' energy efficiency. It was ludicrous that in the age of computers the social security system could not deal flexibly with local variation in benefit payments. The simplified benefit system had been regressive for those in fuel poverty.

Tony Newton repeated that the extra funds (over £300 million) for premium payments were intended for additional targeting of vulnerable groups such as the over-80s and for pensioner premiums for the 60-75 age group. On the issue of flexible targeting of benefits, he said that it would be extremely hard to devise a means of assessing the number of individual homes that would be involved, and in any case it would not be possible to ensure that money paid for heating costs was actually used for that purpose. Brenda Boardman's proposal would imply the need to take away responsibility from the recipient, decide how much would be required for heating and ensure that the appropriate benefit amount was indeed spent on fuel. Moreover, if the target group is taken as eligible under the Home Energy Efficiency Scheme (HEES), it would involve some 7 million homes and individual household assessment of reasonable needs and expenditure. The scale of the task would be administratively impractical, and households would have to be reassessed every time some insulation improvement was made.

Other proposals included the suggestion that the HEES scheme ought to target those people identified as being most in need, such as the 300,000 on the Fuel direct scheme. Tony Newton said that there was no necessary connection between being on Fuel Direct and living in an energy-inefficient home. He doubted whether the DSS had any information that would make a connection between direct deductions for fuel bills with the level of insulation in the households in question. In fact many of those houses might not have an insulation problem at all. He would, however, examine the idea of supplying information on HEES to Fuel Direct claimants.

Turning to the simplification of DSS procedures, it was argued that it had been of more assistance to the Department than to recipients of benefit. The Secretary of State was challenged to make a statement of the Government's ambition about reducing the revenue expenditure of the DSS by promoting extra capital expenditure to cure fuel poverty and address the scale of the problem of improving energy-inefficient housing stock. At current rates it would take a century or two to upgrade the housing stock of the UK.

Tony Newton replied that the simplification of DSS arrangements had helped claimants by cutting delays and improving officers' understanding of the system. He went on to say that there had been transfers of DSS money to help in developing the predecessor of the HEES scheme, and that the legislative vehicle for HEES had been a social security bill: there was thus no question of the DSS not acknowledging the links between social security and energy efficiency issues.

Employment aspects of energy efficiency

Linda P. Taylor
Head of Policy Development
Association for the Conservation of Energy

Introduction

Human activities are changing the composition of the earth's atmosphere, with potentially serious, though as yet unpredictable, consequences for global temperature and climate. The main contribution is the burning of fossil fuels as a source of energy, and it is now almost universally recognised that improving energy efficiency is the quickest and most cost-effective way to combat global warming.

The potential for energy saving is enormous. Government figures suggest that half of the £50 billion spent on energy each year in the UK is wasted, much of it through heat loss from buildings. By investing in measures which pay for themselves in five years or less, the UK could save £10 billion a year from the national fuel bill. Including measures with longer payback periods could save up to half of the national fuel bill (1).

Investment in energy conservation yields positive returns in several different areas, including:

- reductions in atmospheric pollution, including methane, carbon dioxide, sulphur dioxide and nitrogen oxides;
- lower fuel bills for consumers as a result of less energy waste;
- increased comfort, and therefore quality of life, for low-income householders with energy inefficient homes;
- improvement in health, for householders in previously cold, damp and mould-ridden homes.

In addition, investment in energy conservation has positive effects on employment by creating new jobs, in manufacturing, delivering and installing energy efficiency products. These jobs are largely unskilled and semi-skilled, with a large proportion occurring in areas of high unemployment, where many of the buildings requiring remedial work are situated. These local economies would be strengthened, when the new wages and the savings from fuel bills were spent locally, or re-invested by local businesses.

Improving the nation's energy efficiency is not, therefore, simply a matter of individual, but of national and international benefit. Government has a critical role to play in ensuring that emission reduction targets are met, but to date it has analysed neither the employment effects of energy conservation, nor the full range of environmental and societal benefits which such investment offers. The absence of a Government study is all the more surprising given that unemployment is currently over two million, with costs to the Exchequer calculated to be £8,000 per annum per unemployed person (2).

The purpose of this paper is threefold:
- to summarise the findings of relevant studies which have been carried out on the employment creation potential of large-scale investment in energy efficiency and energy conservation;
- to draw broad conclusions from these studies, and
- to suggest areas for discussion.

Several of the studies give detailed analysis of the macroeconomic effects of the investments they propose: to reproduce these is beyond the scope of this briefing, which attempts simply to present the main conclusions of each study.

Methodology
The reports summarised are:

'Jobs and Energy Conservation,' Environmental Resources Ltd. (ERL) for the Association for the Conservation of Energy (ACE), 1983;

'Employment Effects of Energy Conservation Investments in EC Countries,' Fraunhofer Institute for the EC, 1985;

'Less Fuel, More Jobs,' PSI, 1985;

'Too Cold for Comfort,' ERR, 1986;

'Energy Investments for a Stronger Louisiana Economy,'
Economic Research Associates, Eugene, Oregon, May 1991;

'Fuel Poverty: From Cold Homes to Affordable Warmth', B.
Boardman, 1991.

Some of the issues addressed in, or raised by, the reports require
some introduction. The key concepts are given below:

Duration of jobs: the ACE/ERL report uses a 10-year framework,
and each job created is assumed to last 10 years. Other studies refer
to job-years. 10 job-years equals 10 jobs which last one year, 2 jobs
which last five years, or one job which lasts 10 years. This briefing
will therefore refer to job-years, unless specified otherwise.

Energy efficiency measures: each study has chosen a particular
package of energy conservation technologies as the basis for its
calculations. The Fraunhofer Institute includes technologies which
are not normally thought of purely as energy efficiency measures, such
as district heating, solar and biogas technology. The ACE/ERL report,
on the other hand, proposes a package of insulation, controls and
lighting improvements, while the ERR report adds Combined Heat
and Power (CHP) and district heating to the ACE/ERL scenario. Each
study therefore estimates a different part of the total employment
potential, if all cost-effective technologies which reduce overall
energy use were included in the programme. In this respect, therefore,
the results can be regarded as conservative.

Costs: the reports use prices from several different years. In order
to permit rough cost comparisons, all figures have been adjusted to
1990 prices, using the UK Retail Price Index to account for inflation.
The original costs, and the year to which they refer, are given in
parentheses. This is a crude method of updating the figures, as it does
not take account of other changes such as wage levels, productivity or
the cost of materials. It does, however, give a rough basis for
comparison of the different proposals.

Cost per job-year: gross and net costs are given where possible.
Gross cost per job-year is the total cost of the programme divided by
the number of job-years created. Unless otherwise specified, net cost
per job-year refers to the gross cost minus £8,000. This is estimated
to be the cost to the Exchequer of each unemployed person, in benefits
paid and tax revenue lost (3). The ACE/ERL report uses a different
method, whereby it gives a net present value per job-year of 66 percent
of gross costs. Figures for both methods are given, so that comparisons

may be made on the same basis. Clearly, the way in which net costs are defined has a strong impact on the final figures.

Evaluation criteria: significant though the employment effect of energy conservation investment may be, it is important to note that job creation is not the prime motivator for the large-scale programmes discussed here. Total energy savings, emission reductions and improved standards of living are the prime objectives, and job creation is a happy by-product. It may be, for example, that investment in other projects would yield a higher number of jobs for the same expenditure, but without the environmental, financial and societal benefits of investment in energy conservation. It is important, therefore, to account for these additional benefits when evaluating energy efficiency investments.

Direct and indirect effects: as well as the jobs created directly by the programmes, (in installation, administration, and in some cases manufacturing), there are also indirect effects on employment. These indirect effects can be significant, and are described as:

- The multiplier effect. This occurs when those people newly-employed in the conservation programme spend their wages on goods and services made or provided in the UK, so creating more jobs.

- Re-spending effects, when the money saved on fuel bills is spent on other things. Some of the savings are taken up in increased comfort levels, and some invested, so that the overall effect is smaller than the multiplier effect.

- Negative employment effects. These occur if jobs are lost in the supply industry as a result of lower energy demand. The reports which address this issue in detail find that the negative effects are not significant. The ACE/ERL report, for example, suggests that 1,000 - 2,000 jobs at most would be lost in the supply industries, compared with 50,000 - 100,000 jobs created. It does not, however, compare the type or quality of jobs involved.

Accuracy: each of the studies examined stresses that its conclusions are approximate. So many variables are involved, that only a detailed input-output analysis could even approach an accurate prediction of the number of jobs which would be created by a large-scale national programme. The reports are unanimous, however, in their conclusion that the effects on employment would be positive.

The Reports

Jobs and Energy Conservation

Environmental Resources Limited (ERL) for the Association for the Conservation of Energy, 1983

The ACE/ERL report posits a 10-year national energy conservation programme directed at reducing space heating in buildings, in all sectors. It develops two investment scenarios: a Base Case with a very strong economic justification, which it estimates to be well within the capacity of the industry to deliver, and a Maximum Case which, although it involves a much higher level of investment, and is at the limit of what could be delivered, still satisfies minimum economic investment criteria (4).

Both cases choose the following energy conservation measures:
- loft insulation
- cavity wall insulation
- solid wall insulation
- roof/ceiling insulation
- draughtproofing
- heating controls
- lighting efficiency
- double glazing (Maximum Case only).

The choice of these measures was based on Government evidence of the cost-effectiveness of energy conservation measures. With the exception of double-glazing, they offer payback periods of 10 years or under (5). Most offer a payback of 3-5 years.

Base Case:
- total investment: £15.5 bn (£10 bn 1982);
- job-years created: 500,000;
- fuel saving by year ten: £2.2 bn (£1.4 bn 1982) per annum;
- CO_2 saving in year ten: 54 million tonnes;
- CO_2 saving from years one-ten: 264 million tonnes;

Maximum Case:
- total investment: £38 bn (£24.5bn 1982);
- job-years created: 1,223,000;
- fuel saving by year ten: £4.3 bn (£2.8 bn 1982) per annum;
- CO_2 saving in year ten: 108 million tonnes;
- CO_2 saving from years one - ten: 500 million tonnes;

Both Cases:
- gross cost per job-year: £31,000;
- net cost per job-year: £20,000 (net present value 66% - see below).

Both Cases assume that Government investment would meet 75 per cent of the costs, ie £11.6 billion (£7.5bn 1982) for the Base Case and £28.5 billion (£18.4bn 1982) for the Maximum Case. The remaining 25 per cent of costs would be met by consumers themselves. Cost to the Exchequer would therefore be £15,000 per job-year.

This report does not use the same definition of net cost per job-year as the others. Instead it uses a net present value based on a Government study of job creation in development areas (6). This report suggested that in order to take into account the reduction in payment of unemployment and other benefits, and the increased tax revenue and National Insurance contributions resulting from the new jobs, a percentage of the gross direct exchequer cost should be applied, to give net cost per job-year. The rate it suggested was 78 per cent. The ACE/ERL report argues that, if the value of future energy savings in central and local government buildings were included, along with a reduction in fuel bill subsidies to householders receiving State Benefits, the net present value per job would be no more than 66 percent of gross exchequer cost (7).

Roughly two-thirds of the jobs created would come directly from the programme, and the remaining third would come indirectly, from the multiplier and re-spending effects. Of the direct jobs created in the Base Case, 12 per cent would be in manufacturing, 65 per cent in installation and 23 per cent in technical and administrative support. In the Maximum Case the manufacturing share increases to about 16 per cent. The proportion from manufacturing is small, partly because the insulation and controls industries are now highly automated, and partly because of over-capacity due to the decline in the market. This is as much the case in 1991 as it was in 1983 when the report was written, as these industries continue to report a decline in the market.

The report emphasises the need for an adequate record of the current state of the building stock in the UK, but neither estimates how many jobs might be created, nor what the cost of a large-scale energy audit excercise would be.

Net financial savings would begin to accrue to householders from year three of the programme, by which time their 25 per cent contribution to the capital cost would have been repaid in savings on fuel bills. When calculating the effects of re-spending, the ACE/ERL report allows for 25 per cent of the savings to be taken in greater warmth in the home (giving an average temperature increase of just under 4°C). Of the remaining savings, 20 per cent is allocated to investment, and the rest, it is assumed, is spent on other things. The report estimates that just over half of this amount would be spent on goods and services from the UK, and half on imported goods. This means that one third of gross savings would contribute to creating jobs in the UK. At an estimated one job per £23,000 of consumer expenditure (£15,000 1982) this gives cumulative totals of approximately 12,000 jobs in the Base Case and 25,000 in the Maximum Case generated by the re-spending of savings from fuel bills.

The negative employment effects in the fuel industries are estimated as follows:

Industry	Demand reduction	Jobs lost
Gas	Base case: 7 - 9%	200
	Max.case: 12 - 14%	500
Coal	Base case: 2 - 3%	250
	Max.case: 4 - 5%	500
Electricity	Base case: 1.5% (800 MW)	250-400
	Max.case: ' 3.0% (1700 MW) 500-888	
TOTAL	Base case:	700-850
	Max.case:	1,500-1,888
Gross job-years	Base case: 500,000	Max.case: 1,223,000
Net job-years	Base case 499,150	Max.case 1,221,112

These calculations must now be regarded as very approximate, since the electricity industry in particular has been completely restructured, and indeed is beginning to shed thousands of jobs, following privatisation. It may well be that in the new, 'slimmer' generating industry, fewer jobs would be lost due to reduced demand. Of course, were such a programme initiated now, utilities could well be actively participating themselves. This possibility is discussed in greater detail below.

Too Cold for Comfort
Simon Hodgkinson for Earth Resources Research, 1986

This study takes as its starting-point the 10-year programme of insulation work proposed in the ACE/ERL Base Case. On to the £15.5 billion programme it grafts a £46.4 billion (£36bn 1986), 30-year combined heat and power (CHP) programme. It proposes that CHP be established on a city-wide basis in, for example, Sheffield, Newcastle, Edinburgh, London, Belfast and Leicester. Major district heating networks would be constructed in conjunction with the CHP plant, and, as the programme developed, additional schemes would be established in Glasgow, Liverpool and Manchester, as well as in smaller towns such as Plymouth, Basildon, Ipswich and Great Yarmouth.

The report estimates that up to 5 million households and thousands of businesses would eventually be connected to District Heating, and that the savings from this fuel switching would be in the order of 25 million tonnes of coal equivalent, or 6 per cent of total primary UK energy consumption.

The report cites a study by the Combined Heat and Power Association, which suggests that 140,000 job-years of work would be created by introducing CHP/DH in 9 major cities in the UK over 10-15 years, and extrapolates from this a total of 1.4 million job-years which would be created directly from its own programme, with an additional 600,000 job-years created indirectly. In all, it estimates that 2.5 million job-years could be created by the combined CHP/efficiency programme it outlines.

In summary, the 30-year programme results in:

- total programme investment of £62 billion (£46bn 1986);
- total job-years created: 2.5 million;
- total fuel saving: 30 million tonnes of coal equivalent, or £3 billion, per annum;
- CO_2 savings: 80 million tonnes per annum;
- gross cost per job-year: £25,000;
- net cost per job-year: £17,000. *

(* When calculating the net cost per job-year, the report uses a cost to the Exchequer per unemployed person of £5,800 (£4,500 1986) per annum, which gives a net cost per job-year of £19,000. The 1991 figure of £8,000 per unemployed person per annum has been applied here, for the sake of consistency).

Applying the method adopted by the ACE/ERL report, and using a net present value of 66 per cent of the gross exchequer cost to offset savings, gives a net cost of £16,400 per job-year.

Employment effects of energy conservation investments in European Community countries
Fraunhofer Institute for European Commission DG XVII, 1985
This report explores the potential for, and effects of, investment in six technologies in four EC member states: Denmark, France; the UK and the old Federal Republic of Germany (West Germany). The technologies, which the report defines as energy efficiency technologies are:
– district heating (using mostly CHP generation);
– insulation of residential buildings;
– heat exchangers for heat recovery;
– large gas engine-driven heat pumps;
– domestic solar hot-water systems, and
– biogas plants in the agriculture sector.
The stated aim of the report is neither to develop nor to cost an energy efficiency programme, but to analyse the impact of an energy demand policy on production, employment, imports and income in the four countries examined. This it does by means of an input-output analysis based on the 1975 tables for the four countries examined. The economic analysis includes indirect effects, and so gives the net effects on the economy. Although this analysis was carried out on a quantitative basis, the authors stress the inherent inaccuracies of the data, and emphasise a qualitative interpretation, namely that the effects monitored are positive, and that the net impact on employment is almost certainly underestimated.
Findings for the UK, over a 38-year period (1983 - 2020), are summarised below:
– total investment cost: £14.6 billion (£9.4 bn 1982);
– job-years created: 594,000;
– energy savings: 5,146 petajoules (PJ), ie 257 PJ per annum, which is approximately 4%, or £2 billion, of UK energy demand;
– CO_2 savings: 15 - 17 million tonnes per annum;
– gross cost per job-year: £24,500;
– net cost per job-year: £16,500.
An average payback of 4 - 5 years was found for all technologies except heat-pumps, which had a payback of 9 - 11 years. The report finds a very high overall rate of return in the UK compared with the

other countries, with savings projected to exceed investment by a factor of three. The authors conclude that this is mainly because the potential for energy conservation in the UK is largely untapped, a conclusion which remains as valid in 1991 as it was in 1985 (8).

Less Fuel, More Jobs
M. Hillman and A. Bollard, Policy Studies Institute, July 1985
This report presents a summary of the main arguments for improved energy conservation in buildings, and the effects on employment are also examined. The report comments that the ACE study does not include the number of jobs which would be created in energy auditing, the installation of pre-payment meters (the authors suggest a total of 10,000 job-years), or in carrying out remedial work to central heating systems capable of being upgraded. It concludes, therefore, that the ACE/ERL report is very conservative in its estimates of the job creation potential of a national energy efficiency programme.

Hillman and Bollard cite the results of several other studies, most notably the evidence presented to the House of Commons Energy Select Committee Inquiry on Energy Conservation in Building in 1981. The Royal Institute of British Architects, for example, submitted evidence that a £1.5 billion annual programme (£800 million 1980) would generate 60,000 - 70,000 related jobs, over a 20-year programme to insulate one million dwellings annually at a unit cost of £750 (£400 1980) (9). This gives a gross cost of £18,000 per job-year, and net costs of £10,000 per job-year.

Evidence to the same Select Committee enquiry, from a group of companies calling themselves the Group of Eight, came up with broadly similar conclusions; namely that a 20-year national programme including loft and wall insulation, draught-proofing, heating controls and selective double glazing could provide up to 70,000 jobs (10).

The authors point out that these schemes all assume central coordination, and suggest that a market approach might generate higher levels of employment by concentrating on the lowest-cost, highest-return measures such as draught-proofing, loft insulation and heating controls, which are also the most labour-intensive measures.

This argument fails, however, to address the problem of lost opportunities. Lost opportunities occur when only those measures with the highest rate of return are taken, and other options, which would have been cost-effective if carried out at the same time, then become less cost-effective to carry out separately at a later date. Considerable energy savings as well as jobs could be lost through such an approach. Such is the case with the Homes Energy Efficiency Scheme (HEES), which covers the cost of draughtproofing and loft insulation only, for householders in receipt of State Benefits. Buildings which have benefited from this work will have to be re-visited at some point in the future to insulate walls, upgrade heating systems etc.

Energy Investments for a Stronger Louisiana Economy
S. Laitner, Economic Research Associates, Eugene, Oregon, May 1991

This study from the USA has been included for two reasons: it employs input-output analysis, and is very recent. The consultancy which undertook the work claims to be the only organisation in the USA (and, one might safely add, in Europe) to carry out this type of analysis.

It proposes a 10-year energy efficiency investment programme in all sectors of the state of Louisiana, within the following constraints:

Goal:	Save 1,000 megawatts of electricity;
Period covered:	1991-2000 for the investments;
	1991-2010 for the total impact analysis;
Payback allowed:	Five years or less.

The target savings are allocated to each sector, with 30 per cent to come from the residential sector, and 70 per cent from the commercial and industrial sectors. Typical paybacks of just under five years in the residential sector, and three to four years in the industrial and commercial sectors are predicted. An investment of $380 million (£217m) is calculated to be necessary to achieve the required savings of 1,000 megawatts, at $38 million per annum over 10 years.

Expenditure over the 10-year period is translated into constant 1991 dollars, and the different efficiency options are analysed using an input-output model. This evaluates the total job and income impacts (the 'output') which are likely to result from each major change in local spending patterns (the 'inputs').

The positive and negative impacts are then analysed to give the total impacts of the programme. The projected saving to ratepayers is $1.1 billion (£630m) over 20 years. Other benefits which would accrue over a period of 20 years include:

Scenario	Employment	Wage & Salary Income
Business-as-Usual	470 jobs	$12.75 million
Efficiency Investment	1,100 jobs	$22.80 million
Net Gain	630 jobs	$10.05 million
Net 20-year Gain	12,600 job-years	$201.00 million

The gross cost per additional job-year created by this programme is therefore $30,000 (£17,000), not allowing for savings to the Government in unemployment benefits and lost tax revenue (11).

Fuel Poverty: From Cold Homes to Affordable Warmth
B. Boardman, Belhaven Press, 1991

This study's central thesis is that fuel poverty is a direct result of hopelessly energy inefficient dwellings, and, as part of the solution, it proposes a national energy efficiency programme for the domestic sector. This programme, dubbed Programme for Affordable Warmth (PAW), proposes to bring an estimated 6.6 million homes up to the thermal insulation levels required by the 1990 Building Regulations, at an average cost of £2,500 per home (based on costings for bulk work by the South London Consortium) (12). The investment required is £16.5 billion, at a rate of £1.25 billion a year between 1992 and 2005. Boardman estimates that around 970,000 job-years would be generated, at a gross cost of £17,000, and a net cost of £9,000 per job-year.

Summary

The findings of the reports are summarised in the table below, with the caveat that they were carried out in several different economic climates (and countries), using different packages of measures, and different methodologies, over different time-scales, and therefore some distortion will have occurred by translating the results into the same terms.

Study	Investment (£bn, gross)	Fuel Savings (£bn / p.a.)	Net Cost per Job-Year*	Job-Years Created	Comments
ACE/ERL	15.5	2.2	23,000	500,000	10-yr programme
	38.0	4.3	23,000	1,223,000	20-yr effects
ERR	62.0	3.0	17,000	2,500,000	30-yr programme
Fraunhofer	14.6 **	2.0	16,500	594,000	38-yr programme
Louisiana	0.22	0.03	9,300	12,600	10-yr programme
Boardman	16.5	n.a.	9,000	970,000	13-yr programme

Notes: * *Calculated for all reports as total investment over number of job-years created, minus £8,000 (including the Louisiana study), so that all figures may be compared on the same basis.*
** *This figure includes savings made in operation and maintenance of plant and equipment, but not fuel savings.*

The above table gives a very simplified picture of some of the returns from the programmes proposed. Estimates of the net cost per job-year created varies significantly, but it is interesting to note that the costs per job-year are lowest in the more recent reports.

How could large-scale programmes, such as those outlined above, best be delivered in the UK?

Issues for discussion
Funding

Investment on the scale and over the duration suggested by the reports examined raises several questions. Where is the funding to come from? What will be the effects on the economy of such a programme? Large-scale public spending in a time of economic growth could have undesirable inflationary impacts, while in a recession public funds are in short supply. There are several ways in which to raise revenue for an energy efficiency programme, without negative impacts on the economy, for example:

A carbon and/or energy tax. The European Commission has proposed the introduction of a tax, based partly on all fuels except renewables, and partly on the carbon content of each fuel. The revenue from a tax set at $10/barrel of oil-equivalent would raise £7 billion per annum in the UK (13).

A gas and electricity levy. The Netherlands has introduced a levy of between 0.5 and 2 percent (the rate is set by each utility), which is used by the gas and electricity distributors to fund rebates for a variety of energy efficiency measures;the 'E' factor which the Office of Gas Regulation (Ofgas) has recently introduced into the gas pricing formula. This allows the gas supplier to recoup the cost of energy efficiency programmes directly through the pricing formula. Investments of £90 million a year would cause an average increase of 10p per consumer of gas per week. The same formula could be applied to the electricity pricing mechanism.

Reallocation of public spending. Reallocation is always controversial, as different priorities compete for the same funds, but at least a percentage of the investment could be found in this way.

New public spending. The Government announced an increase of £11 billion in public spending in its Autumn Statement, demonstrating that relatively large expenditure can be justified: the difficulty would arise in securing a committment of funding over a period of ten years.

Delivery

There are several different ways in which a large-scale energy efficiency programme could be delivered. For example:

The Homes Energy Efficiency Scheme (HEES) is the nearest the UK has come to a national energy efficiency programme. With a budget of £24 million in 1991, and £40 million for 1992/93 (compared with the average £1bn/year called for by the reports reviewed above), it provides grants towards the cost of roof insulation and draughtproofing for householders in receipt of State Benefits. At its present rate of 200,000 houses a year, it will take 30 years to cover the 6 million eligible households (or 24 years, if this year's higher budget level is maintained).

The HEES has been criticised for being too restricted in terms of the measures it offers. In addition, many of the organisations involved have not yet succeeded in establishing a high standard of work (14).

Alternatively, the responsibility for programme administration could be given to *Local Authorities*. A study by the Institute of Employment has identified Local Authorities as a prime vehicle for local jobs plans, since they are already major employers (of 2.5m people in total), and well placed to match the supply of, and demand for, labour (15).

The gas and electricity utilities are another possibility. A national energy efficiency programme involving the spending of an estimated £4 billion over 10 years has recently been introduced in the Netherlands. The Dutch utilities are central to this programme: they administer the grants available to householders, and offer energy efficiency services. In the UK, British Gas has been closely involved with community insulation projects for some time, and many of the electricity distribution companies offer energy efficiency services, including insulation work. It is therefore perfectly conceivable that the utilities might be a vehicle for a major energy efficiency programme in the UK.

Quality control and training

Quality of work and training for the job must be a major component of a larger programme. The effectiveness of energy efficiency measures is significantly affected by how well it is installed. Studies by Rockwool, for example, have shown that if roof insulation is laid askew by even 10mm (half an inch) between each set of joists, 15 percent of potential savings will be lost. If the rolls of insulation are laid across, rather than between, the joists, losses of up to 50 percent can be expected (16). Other problems, notably condensation and freezing pipes and tanks, can result from poor installation. National Vocational Qualifications (NVQs) in insulation and energy awareness have been developed by Neighbourhood Energy Action, and could therefore be used as a basis for training. Qualified personnel and proven standards of installation could be a minimum requirement for contractors on a large-scale programme.

Another area in which training is important, and which has implications for job-creation, is advice for householders. Improving the thermal property of the building has a significant effect on comfort levels, but especially where central heating is concerned, a proper understanding of how to regulate and monitor the system is essential. Householders need to understand how to set timers, how to choose between hot-water and heating programmes and how to make the most of thermostatic radiator valves. In addition, many householders are unclear about the relation between meter readings and their fuel bills, and so may mistakenly believe, for example, that gas is more expensive than electricity. Brenda Boardman distinguishes between the provision of information (of general relevance, distributed via leaflets, brochures etc), and advice, which is specific to an individual household, and delivered in person, preferably in the home (17). Advice is considerably more likely to effect changes in behaviour than information alone, but is much more expensive to deliver.

Focus

The priority of a large-scale investment programme must be to save energy, rather than to create jobs. The positive effect on employment is only one of several benefits arising from accelerated investment in energy conservation. Without a clear focus, a large-scale programme risks becoming yet another job-creation scheme, with negative consequences for the effectiveness of the programme. As an example, the Neighbourhood Energy Action (NEA) community insulation projects, which have installed draughtproofing and loft insulation in nearly 1 million homes in the last ten years, have suffered from conflicting priorities. NEA's main priority has been energy conservation and the eradication of fuel poverty, but the Government funding programmes on which it has had to rely have been primarily concerned with job creation. Some of the results of this arrangement have been the imposition of part-time work, low wages and high turnover of staff, all of which have adversely affected productivity and quality.

Beneficiaries

Who should benefit from the kind of programme proposed above? The seven million households identified by Brenda Boardman as suffering from fuel poverty and very energy inefficient homes are an obvious priority. Yet the potential to save energy is higher in households that can afford to heat their homes, and therefore to waste energy. For maximum impact, the aim of a national energy efficiency programme should be to bring all buildings up to a given level of thermal efficiency, such as the 1990 Building Regulations.

Public sector and Housing Association property would be the easiest to reach, but many of the neediest households live in their own, or in privately rented, accommodation. A combination of incentives, encouragement and regulation may be required to penetrate the private sector to a significant degree. An important role could be played by the proposed home energy labelling scheme, whereby every home sold would be required to have an energy audit, which would grant a certificate giving details of energy use, running costs, etc. Grants could be made towards the cost of the audit, with additional grants being provided, on a sliding scale based on disposable income, towards the cost of remedial work. If this could be linked to bulk work already being carried out in public sector housing, costs would be kept low.

Conclusions

Despite the different programmes and measures they propose, the reports reviewed above come to broadly similar conclusions, namely that investment in energy conservation yields positive returns, including:

- improved energy services, through lower fuel bills and increased thermal efficiency;
- increased comfort, and therefore quality of life, for low-income householders with inefficient homes (with attendant improvements to health);
- reductions in atmospheric pollution;
- a positive impact on employment, and
- a strengthening of local economies.

Where they do differ significantly is in their costings. Some of these differences may result from the measures chosen, or from the economic circumstances relevant at the time of writing, as well as from the depth of analysis involved in each report.

It is clear that a national programme to improve the thermal efficiency of the UK housing stock would be a major undertaking in terms of organisation, funding and delivery. A much clearer and more up-to-date picture of the macroeconomic effects involved would contribute greatly to the planning process for such a programme.

References

1. The Government defines cost-effective energy efficiency measures as those yielding a payback at current costs of under 5 years. By contrast, many Government investments in public services - power stations, for example - are allowed a much longer payback: 20 years is standard.
2. Institute of Employment figures, quoted in *Unemployment: solutions for the 90s,* Campaign for Work, June 1991.
3. Institute of Employment, ibid.
4. The minimum investment criteria are that the payback period should be shorter than the lifetime of the measure.
5. Evidence quoted in the Fifth Report from the Select Committee on Energy: *Energy Conservation in Buildings,* Vol II, Appendix 10, HMSO, London 1982.
6. *Measuring the Effects and Costs of Regional Incentives,* Government Economic Service, Working Paper No 32, Dept of Industry, Feb. 1980.
7. Heating Additions, previously payable to certain categories of people receiving State Benefits, were abolished on 1 April, 1988.
8. Even solar water systems are found to be cost-effective in the UK, mainly based on very poor substituted hot-water devices of extremely low efficiency. Paybacks in Denmark, for example, are from 28-50 years.
9. RIBA, Chartered Institute of Building Services and the Royal Institution of Chartered Surveyors, *Memorandum of Evidence to the House of Commons Select Committee on Energy,* 1981, HC (1980-81) 253-iv.
10. Group of Eight, *Energy Conservation in Construction with Particular Reference to Employment,* HC (1980-81) 352-vii.
11. At an exchange rate of $1.75/£1.
12. Figures from Newark and Sherwood District Council support this costing (personal communication from the Chief Architect,

Newark and Sherwood DC). A typical package of energy efficiency measures includes:
- 100mm loft insulation added to existing insulation, plus tanks/pipes/hatch insulated;
- cavity wall insulation;
- upgrading of central heating, including: 3 additional radiators (in bedrooms), thermostatic radiator valves for all radiators, new programmer, hot-water cylinder thermostat, room thermostat and reflectors behind radiators.

The total average price, including labour and administration costs, is £750 per dwelling. Where no central heating system exists, or the existing system must be replaced, the cost is £1,900 per dwelling (including gas condensing boiler). At an average cost of £1,325, an investment programme of £1bn a year would allow over 750,000 dwellings a year to be upgraded to these standards. Replacement windows would increase the cost, but should be included where appropriate.

13. From the European Commission, *A Community Strategy to limit Carbon Dioxide emissions and to improve Energy Efficiency,* Communication from the Commission to the Council, October 1991.
14. Failure rates of up to 51% have been reported for the first six months of the scheme by the administering Energy Action Grants Agency (EAGA).
15. Charter for Jobs, *Economic Report,* Vol 2, No.10, Aug. 1987.
16. Personal communication from A.B. Hemson, Rockwool plc.
17. Brenda Boardman, *Aspects of energy literacy and public awareness,* seminar presentation, 'Energy Education in the UK', Cranfield, 1st October 1991.

Summary of seminar on employment aspects of energy efficiency

Chaired by: Sir John Cassels, Director of the Paul Hamlyn Foundation
National Commission on Education
Speaker: Frank Dobson MP, Labour Party Energy Spokesman

In his introduction to the seminar, Frank Dobson argued that the UK had more scope for major energy efficiency savings than many other developed countries because we squander so much energy and start from a very low base. Thus any UK investment in energy efficiency should show large returns. But as well as wasting energy the UK also managed 'to freeze more old people to death in winter than any other country in Northern Europe', and had around 7 million families living in fuel poverty. So the scope for improving energy efficiency is huge, but the issue has not been given the priority it deserves. In order to get the issue further up the policy agenda a Labour government would establish an Energy Efficiency Agency with a degree of independence from Government, whose sole task it would be to promote energy efficiency and conservation, staffed by people prepared to devote a significant part of their careers to the development and implementation of a national energy efficiency programme.

The first priority of a Labour energy policy would be energy saving rather than energy sales, a reversal of established policies which would for the first time put customers' needs before producers' interests. Hence the need for an agency to promote efficiency and conservation: all other bodies in the energy producing field are dedicated to promoting the sale of energy. There was a long term need to adjust the tariff structures of energy producers and develop their role as providers of services (eg warmth) rather than simply sellers of fuel and power. A Labour administration would wish the utilities to promote energy saving in earnest and not simply pay lip service to it.

The first step would be to 'kickstart' the utilities by changing their licenses to insist on investment in the insulation of a designated number of homes in all regions of the UK. Gas and electricity producers had a great deal to offer to consumers in this respect. VAT would also be reduced on energy-saving equipment, and rapid progress would be made on energy labelling of products. These measures would save energy, help the environment and balance of payments, husband fuel reserves and save consumers money.

However, Frank Dobson went on to point out that there was a further benefit from such a package of measures, in terms of employment. The clear message from the studies summarised in Linda Taylor's paper is that energy efficiency policies could help to create large numbers of jobs. The key questions were: what kind of jobs would be created, where, at what cost and when? He noted that the emphasis in the studies cited is mainly on semi-skilled and unskilled jobs in areas such as loft insulation, targeted on areas of highest unemployment where housing conditions are poorest. But, he argued, the so-called semi-skilled tasks usually called for training and a painstaking approach, especially in the light of complaints over the poor standards of insulation work carried out to date. Training standards and high quality of work would be priorities for the proposed Agency. Without high standards of training the work would be slapdash and the initiative would lose credibility.

Frank Dobson maintained that over the coming decade there would be growing demand in the developed world for energy-efficient plant and machinery, leading to demand for high quality high-tech work in design, installation, operation, maintenance and repair. Standards would be tightened such that sale or installation of energy inefficient equipment would be proscribed. This trend was inevitable: however, it was not inevitable that the design and manufacture of highly energy efficient technologies would be carried out in the UK. The only way to ensure that the UK gets the jobs that would be generated was to raise quality standards immediately in industry. A Labour government, in consultation with industry, would set standards for quality and training and seek to boost R&D in order to promote excellence in UK industry. He argued that we could not afford not to raise standards: low standards would lead eventually to loss of jobs, export markets and domestic markets.

If new market opportunities were to be seized there needed to be more collaboration between the gas and electricity suppliers and industry in the development and marketing of energy-efficient equipment. Unfortunately in his view the utilities had done a poor job to date in providing good advice to consumers on energy efficiency, training showroom staff and demonstrating commitment to the cause. In addition, the recent concentration by government in publicity campaigns had been an ineffectual policy – he argued that they simply gave the impression of more activity going on than was actually the case. Half of the Government Departments increased their spending on energy in 1991 and most did not meet the Government's criteria for investment in energy saving in relation to the sums spent. So, he concluded, there was a considerable way to go in Government, which had tended to tell us to 'do as I say, not as I do'.

In his view, the privatisation of the energy utilities had led to an exclusive focus in their reports on selling energy and expanding sales to maximise profitability. Every part of the privatised energy sector made more money the more units it sold, which was a fundamental flaw. The same was true, he argued, of the European Energy Charter, which in its present form was solely concerned with creating a bigger energy market.

Discussion
We highlight below the key points relating to employment and training issues raised in the course of a wide-ranging discussion of the overall package of measures proposed by Frank Dobson.

Part of the job-generating potential of increased energy efficiency investments lies in the need for improved information and advice services, as noted in all of the seminar discussions. Frank Dobson was asked about the scope for provision of independent advice centres for the consumer, covering all forms of fuel and not tied to particular generating companies. He took the view that there was an undoubted need for independent advice and information services, but that it was not evident that a single agency to provide such services was a good idea. Different groups of consumers had different information needs; moreover, the large industrial users of energy had sufficient expertise and influence to obtain the information they needed from the utilities, and so domestic consumers should not be expected to pay for advice services that catered to big business as well as households and smaller firms. In general he thought that it would be preferable to develop new services using existing outlets rather than establishing new ones, provided that consumers felt that the information and advice received was unbiased. Given the present structure of the generating industry and the competition between privatised gas and electricity suppliers, it might be difficult for the public to perceive the utilities as sources of impartial advice and information. It was possible that joint gas and electricity showrooms could be set up at some stage, and perhaps there could be a role for an Energy Efficiency Agency in monitoring the quality of information and advice services to the public. It was acknowledged in relation to this point that British Gas was already required to provde independent energy advice as part of its agreement with the regulating authority Ofgas, and was providing suitable training to showroom staff to standards set by Neighbourhood Energy Action.

Concerns were expressed by the TUC about the potential for increased unemployment resulting from the displacement of workers due to energy efficiency improvements in energy-intensive industries. What policies would be needed for retraining and redeployment? And what measures would be needed for raising awareness in the workplace of the need for energy saving? Frank Dobson replied that if the utilities did all they need to do to improve efficiency they would find that they needed to retain engineering and sales staff rather than shed them. If some jobs were lost on the generating side there should be good potential for displaced people with experience to get new ones within the industry. Energy efficiency in itself, he asserted, would not cost jobs, since the utilities would need to commit resources to new investment and make use of many skilled people. He noted that there would be a growth in high technology jobs from the need for the development and diffusion of more sophisticated instrumentation to monitor large-scale boiler systems. As for raising awareness of energy saving in the workplace, he agreed that there was a need for more action but said that there was also great enthusiasm in many workplaces for energy efficiency measures, with some companies receiving valuable advice from their own staff on areas in which savings could be made.

Attention was also drawn to the fact that Neighbourhood Energy Action had been centrally involved in training and securing employment in home insulation work for large numbers of people for years, but those people were commonly on short-term training programmes and their skills were therefore lost to NEA's network when those initiatives came to an end. This meant that there are many people with some energy efficiency work experience now unemployed again and ready for work. If the HEES scheme and other ventures had substantially more resources there would be scope for worthwhile jobs for these people to fill. What extra resources could be committed to setting up more comprehensive programmes for training and employment in insulation?

Frank Dobson replied that initially, pending further EC policy developments on energy taxes, a possible source of funds would be the gas and electricity industries themselves. They would not have to do all the insulation work themselves and would be free to meet their targets by drawing on existing expertise among people who have been through previous training and employment schemes. In fact, given the quotas that would be set for insulating properties in the poorest condition, they would need to cooperate with current schemes that assist low-income households, and use the same routes for providing improvements. A key issue was the need for the job of loft insulators to be seen as important and of high quality, and for insulation jobs to be well paid, rather than regarded as a field with rapid staff turnover and low pay.